Calcium Signalling

The Practical Approach Series

Related **Practical Approach** Series Titles

Protein Ligand Interactions: structure and spectroscopy
Protein Ligand Interactions: hydrodynamic and calorimetry
Cytoskeleton: signalling and cell regulation
Protein Phosphorylation 2/e
Signal Transduction 2/e
Steroid/Nuclear Receptor Superfamily
Growth Factors and Receptors
Signalling by Inositides
Subcellular Fractionation
Ion Channels

Please see the **Practical Approach** series website at
http://www.oup.co.uk/pas
for full contents lists of all Practical Approach titles.

Calcium Signalling

Second Edition

A Practical Approach

Edited by

Alexei Tepikin

Physiological Laboratory

The University of Liverpool, U.K.

Great Clarendon Street, Oxford OX2 6DP

Oxford University Press is a department of the University of Oxford.
It furthers the University's objective of excellence in research,
scholarship, and education by publishing worldwide in

Oxford New York

Athens Auckland Bangkok Bogotá Buenos Aires Calcutta Cape Town
Chennai Dar es Salaam Delhi Florence Hong Kong Istanbul Karachi
Kuala Lumpur Madrid Melbourne Mexico City Mumbai Nairobi Paris
São Paulo Singapore Taipei Tokyo Toronto Warsaw

with associated companies in Berlin Ibadan

Published in the United States by Oxford University Press Inc., New York

First Edition First published 1991
Second Edition First published 2001

British Library Cataloguing in Publication Data
Data available

Library of Congress Cataloging in Publication Data
Calcium signalling : a practical approach / edited by Alexei Tepikin—2nd ed.
—(The practical approach series ; 246)
Rev. ed. of: Cellular calcium / edited by James G. McCormack and P. H. Cobbold. c1991.
Includes bibliographical references and index.
1. Calcium—Physiological effect—Research—Methodology. 2. Calcium—Analysis. 3.
Second messengers (Biochemistry)—Research—Methodology. 4. Cellular signal
transduction—Research—Methodology. I. Tepikin, Alexei. II. Cellular calcium. III. Series.
[DNLM: 1. Calcium—analysis. 2. Calcium Signaling. 3. Calcium Channels. 4.
Fluorescent Dyes. 5. Photometry—methods. QV 276 C144425 2001]
QP535.C2 C46 2001 572′.516—dc21 00-060683

1 3 5 7 9 10 8 6 4 2

ISBN 0-19-963848-9 (Hbk.)
ISBN 0-19-963847-0 (Pbk.)

Typeset in Swift by Footnote Graphics, Warminster, Wilts
Printed in Great Britain on acid-free paper
by The Bath Press, Bath, Avon

Preface

The previous version of the practical approach book on calcium signalling (edited by McCormack and Cobbold) is now ageing very gracefully.[a] I have seen the copies of the previous edition in many laboratories slowly grinded to dust by inquisitive hands of Ph.D. students and post-doctoral scientists. I hope that the books of this new edition will meet the same glorious end in a few years time.

This volume is mainly about the applications of fluorescent and bioluminescent techniques for studies of calcium signalling (Chapters 1, 2, 4, 6, 7, 9, and 11), and calcium-dependent processes like single cell gene expression (Chapter 8) and mitochondrial metabolism (Chapter 5). This book also contains a chapter on artefacts and problems of fluorescent calcium measurements (Chapter 3) and a description of the pharmacological approach to calcium signalling (Chapter 10). The contents of the book signify the maturation and branching of the technologies adopted for calcium measurements or developed in the calcium signalling field itself.

It was approximately 15 years ago when scores of scientists were recruited to calcium studies by the major technological development—a generation of new calcium indicators (quin-2, fura-2, Indo-1) produced by Roger Tsien and his colleagues. Looking down the microscope onto cells, which shine like jewels on black velvet and wink at you about their calcium signalling, is highly addictive (this addiction can easily last more than 15 years, so be careful). Another development that galvanized the calcium research field, at approximately the same time, was the discovery by Michael Berridge and colleagues of the messenger (calcium releasing) role of IP_3. The calcium signalling field refused to decline after these initial explosions, it continues to grow both in terms of generation of conceptual knowledge and in production of new techniques valuable for cell physiology. The recent and potentially very powerful novel technology in calcium measurement (chameleons) is described in Chapter 1. The majority of the chapters are written by scientists from different countries who during their recent post-doctoral years actually developed the techniques and applications described in their chapters.

---

[a] Now available on CD.

It is difficult to recognize the scientific and technological revolution when you live at the time when it happens; let us hope that we are still living through one.

Liverpool
October 2000

A. V. T.

Contents

Protocol list

Abbreviations

ACh	acetylcholine
cADPR	cyclic ADP ribose
AM (as suffix)	acetoxymethyl ester
ANT	adenine nucleotide translocase
AOD	acusto-optical deflector
asc	ascorbic acid
ASW	artificial sea water
ATPase	ATP consuming enzyme
BAPTA	1,2 bis (*o*-amino phenoxy)ethane-*N*,*N*,*N*′,*N*′-tetraacetic acid
BFP	blue mutant of GFP
BP	band pass filter
BSP	bromosulfopthalein
$[Ca^{2+}]_c$	concentration of free cytoplasmic Ca^{2+} ions
CaM	calmodulin
CCCP	carbonyl cyanide *m*-chlorophenyl hydrazone
CCD	charge coupled device
CCK	cholecystokinin
CICR	Ca^{2+}-induced Ca^{2+} release
CN^-	cyanide ion
CRsig	calreticulin signal sequence
CsA	cyclosporin A
cypD	cyclophylin D
DAPPAC	diamino pentane pentammic acid
DC	dichroic mirror
DEDTC	diethyldithiocarbamic acid
DMEM	Dulbecco's modified Eagle's medium
DMSO	dimethyl sulfoxide
DNP	2,4-dinitrophenol
2,2′-DPD	2,2′-dipyridyl
$\Delta\Psi_m$	mitochondrial inner membrane potential
ECFP	enhanced cyan mutant of GFP
EDTA	ethylenediaminetetraacetic acid
EGF	epidermal growth factor

EGTA	ethylenglycol-bis-(2-aminoethylether)-*N*,*N*,*N*′,*N*′-tetraacetic acid
ER	endoplasmic reticulum
EtOH	ethanol
EYFP	enhanced yellow mutant of GFP
FACS	fluorescence activated cell sorting
FAD	oxidised flavin adenine dinucleotide
$FADH_2$	reduced flavin adenine nucleotide
FCCP	carbonyl cyanide *p*-trifluoromethoxy-phenylhydrazone
FDG	fluorescein-β-di-galactopyranoside
F_1F_0-ATPase	F_1F_0-ATP-synthase
FRET	fluorescence resonance energy transfer
GDP	guanosine 5′-diphosphate
GFP	green fluorescent protein
cGMP	guanosine 3′:5′-cyclic monophosphate
GPD	glycerol phosphate dehydrogenase
GR	glucocorticoid receptor
HBS	Hepes-buffered saline
HBSS	Hanks' balanced salt solution
HeNe (laser)	helium–neon (laser)
Hepes	4-(2-hydroxyethyl)-1-piperazine ethanesulfonic acid
ICP-MS	inductively coupled plasma mass spectrometry
I_{CRAC}	calcium release-activated calcium current
IP_3	inositol 1,4,5-trisphosphate
IP_3R	IP_3 receptor
IPRL	intact perfused rat liver
JC-1	5,5′,6,6′-tetrachloro-1,1′,3,3′-tetraethylbenzamidazolocabocyanine
K_d	dissociation constant
KRB	Krebs–Ringer modified buffer
kz	Kozak consensus sequence
LP	long pass filter
LUT	look-up table
MDR	multidrug resistance transporter
MgG	Magnesium Green
MOPS	3-(*N*-morpholino)propanesulfonic acid
MPME	multi-photon molecular excitation
MTG	mito tracker green
mvCsA	methyl-valine cyclosporin A
NA	numerical aperture
NAADP	nicotinate adenine dinucleotide phosphate
NAD	nicotinamide adenine dinucleotide
NADH	reduced nicotinamide adenine dinucleotide
NADP	nicotinamide adenine dinucleotide phosphate
NADPH	reduced nicotinamide adenine dinucleotide phosphate
NGD	nicotinamide guanine dinucleotide
NGF	nerve growth factor
NLS	nuclear localization signal

NMDA	*N*-methyl-D-aspartate
NMR	nuclear magnetic resonance
PBS	phosphate-buffered saline
PCR	polymerase chain reaction
PDL	poly-D-lysine
PMT	photomultiplier tube
PSF	point-spread function
PTP	permeability transition pore
RFU	relative fluorescence units
Rh123	rhodamine-123
RT	room temperature
Ru360	membrane permeable analogue of ruthenium red
RuR	ruthenium red
RyR	ryanodine receptor
SBTI	soya bean trypsin inhibitor
SERCA	sarcoplasmic–endoplasmic reticulum Ca^{2+} ATPase
SOD	superoxide dismutase
SR	sarcoplasmic reticulum
ST	sialiltransferase
tBuBHQ	2,5-di(tert-butyl)-1,4-benzohydroquinine
TCA cycle	tricarboxylic acid cycle
TFA	trifluoroacetic acid
TMB-8	3,4,5-trimethoxybenzoic acid 8-(diethylamino)-ocyl ester
TMPD	tetramethyl-*p*-phenylenediamine
TMRE	tetramethyl-rhodamine ethyl ester
TMRM	tetramethyl-rhodamine methyl ester
TPEN	*N,N,N′,N′*-tetrakis (2-pyridylmethyl)ethylenediamine
TPP^+	tetraphenyl phosphonium
TRITC	tetramethylrhodamine isothiocyanate

Part One

New probes, new instruments, new methods

Chapter 1

Cameleons as cytosolic and intra-organellar calcium probes

Atsushi Miyawaki and Hideaki Mizuno
Laboratory for Cell Function and Dynamics, ATDC, BSI Riken 2-1 Hirosawa, Wako-city, Saitama 351-0198, Japan.

Juan Llopis* and Roger Y. Tsien
Howard Hughes Medical Institute, University of California, San Diego, La Jolla, CA 92093-0647, USA.

Kees Jalink
Department of Cell Biology H1, The Netherlands Cancer Institute, Plesmanlaan 121, 1066 CX Amsterdam, The Netherlands.

1 Introduction

Cameleons are genetically encoded fluorescent indicators for Ca^{2+} in which a short wavelength mutant of green fluorescent protein (GFP), calmodulin (CaM), a CaM-binding peptide, and a long wavelength mutant of GFP are tandemly fused (1). They offer considerable promise for monitoring Ca^{2+} in previously unexplored organisms, tissues, organelles, and submicroscopic environments. This chapter describes how best to use the indicators for more dynamic and quantitative Ca^{2+} measurements in cytosol, nucleoplasm, and the lumen of endoplasmic reticulum of cultured mammalian cells.

2 Cameleons as new generation indicators for Ca^{2+}

2.1 Advantages of cameleons over synthetic fluorescent calcium chelators and photoprotein aequorin

Cytosolic and organellar free Ca^{2+} concentrations show the most dramatic spatial and temporal fluctuations of any signal transduction pathway. Ca^{2+} signals are most often measured using synthetic fluorescent chelators (2, and Chapters 2 and 9) or the photoprotein aequorin (3, and Chapter 4). The chelators are easily imaged, but they:

(a) Are hard to target precisely to specific intracellular locations.

*Current address: Facultad de Medicina, Universidad de Castilla - La Mancha, Edifico Benjamin Palencia Avda., España s/n 02071 Albacete, SPAIN

(b) Can be perturbed by cellular components.

(c) Gradually leak out of cells, particularly at warmer temperatures.

The non-invasive method for loading chelators via their AM (acetoxymethyl) esters:

(d) Often deposits the indicators in intracellular organelles (see Chapter 6) to unpredictable extents.

(e) Is frequently difficult in thick tissues, intact organisms, or non-mammalian cells.

Genetically expressed aequorin is easily targeted, but it:

(a) Requires the incorporation of the cofactor coelenterazine to emit light.

(b) Is irreversibly destroyed by Ca^{2+} ion, so that its light output depends on the entire past history of Ca^{2+} exposure.

(c) Is very difficult to image because its luminescence produces much less photons than fluorescence.

To combine the advantages of molecular biological targeting and fluorescence read-out, the fluorescent protein indicators for Ca^{2+}, cameleons, have been invented.

2.2 Evolution of cameleons

Cameleons are chimeric proteins consisting of a blue or cyan mutant of green fluorescent protein (GFP), calmodulin (CaM), a glycylglycine linker (4), the CaM-binding peptide of myosin light chain kinase (M13) (5), and a green or yellow version of GFP. Ca^{2+} binding to the CaM causes intramolecular CaM binding to M13. The resulting change from an extended to a more compact conformation increases the efficiency of fluorescence resonance energy transfer (FRET) between the shorter to the longer wavelength mutant GFP. To obtain adequate expression and brightness of the mutant GFPs in mammalian cells, enhanced genes with mammalian codon usage and mutations for improved folding at 37 °C were developed (6). Also the blue mutant (BFP) proved to be the dimmest and most bleachable of the GFPs. It also required ultraviolet excitation, which is potentially injurious, excites the most cellular autofluorescence, and could interfere with the use of caged compounds. Therefore enhanced cyan and yellow mutants ECFP and EYFP have been substituted for the original blue and green mutants respectively (1), to make 'yellow cameleons' (see *Figure 1*). Mutations have been also introduced into the Ca^{2+}-binding loops of CaM to tune the affinity of yellow cameleons for Ca^{2+} (1, 7). Despite the considerable promise of yellow cameleons, they still have problems that need amelioration (8). One of the problems is that EYFP is quenched by acidification. This problem perturbed the signals of yellow cameleons, mimicking a decrease in $[Ca^{2+}]$ when the cellular environment acidified. The pH-sensitivity of yellow cameleons have been greatly reduced by introducing mutations V68L and Q69K into EYFP

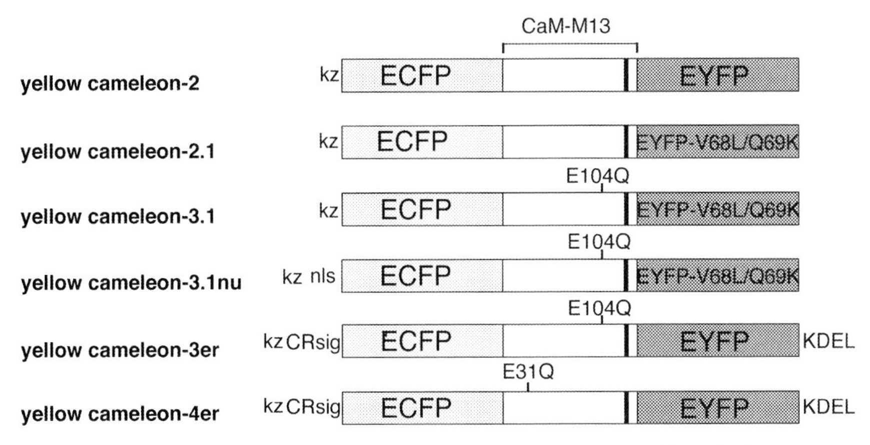

Figure 1 Schematic structures of yellow cameleons for expression and imaging in mammalian cells. E104Q and E31Q, mutations of the conserved bidentate glutamic acid to glutamine at position 12 of the third and first Ca^{2+}-binding loop of CaM, respectively. CRsig, calreticulin signal sequence, MLLSVPLLLGLLGLAAAD; nls, nuclear localization signal, PKKKRKVEDA; KDEL, ER retention signal; kz, Kozak consensus sequence for optimal translational initiation in mammalian cells.

(EYFP–V68L/Q69K). The improved yellow cameleons (yellow cameleon-2.1 and -3.1) permit Ca^{2+} measurements despite significant cytosolic or nuclear acidification (9).

3 Measurements of $[Ca^{2+}]_c$ (cytosolic Ca^{2+} concentration) using improved yellow cameleons

3.1 Improved yellow cameleons (yellow cameleons-2.1 and -3.1)

Yellow cameleon-2.1 gives about a twofold increase in emission ratio between zero and saturating Ca^{2+} (see *Figure 2*). Yellow cameleon-2.1 shows a biphasic Ca^{2+} dependency (apparent dissociation constant K'_d, 100 nM and 4.3 μM; Hill coefficient n, 1.8 and 0.6) and is most responsive near basal cytosolic Ca^{2+} concentrations. Yellow cameleon-3.1 displays a monophasic Ca^{2+} response curve (K'_d, 1.5 μM; n, 1.1) and should be helpful in quantifying relatively large $[Ca^{2+}]_i$ transients.

Figure 3 compares the emission ratio responses of yellow cameleon-2.1 (A) and -3.1 (B) to a supramaximal dose of histamine (0.1 mM) in HeLa cells. The initial peak and subsequent plateau in $[Ca^{2+}]_i$ nearly saturated the yellow cameleon-2.1 response, though in one cell a sustained oscillation with a mean frequency of 0.05 Hz was superimposed upon the plateau. Application of cyproheptadine, a histamine antagonist, caused a large fall in $[Ca^{2+}]_i$ to previous resting values. By contrast, yellow cameleon-3.1 reported apparently much sharper spikes of $[Ca^{2+}]_i$. All the spikes except for the very first reached only about halfway

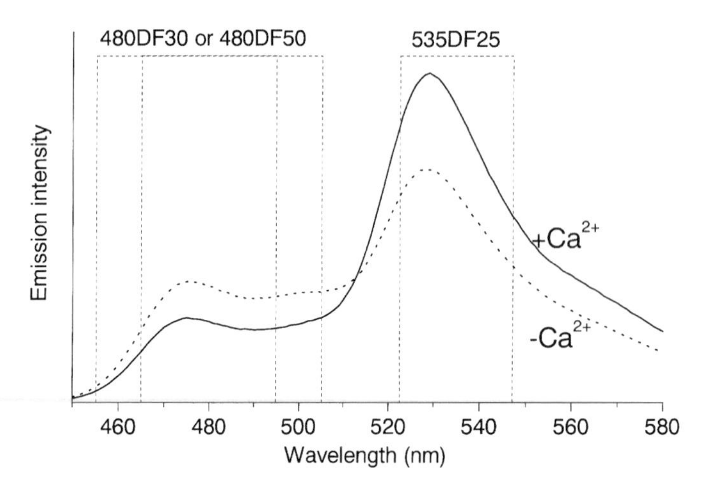

Figure 2 Emission spectra of yellow cameleon-3.1 (excited at 432 nm) with 100 μM EGTA (solid line) and 100 μM $CaCl_2$ (dotted line) at pH 7.3, along with the band pass filters to form the images for the ratio: 480DF30 (480 ± 15) or 480DF50 (480 ± 25) as donor channel, and 535DF25 (535 ± 12.5) as FRET channel.

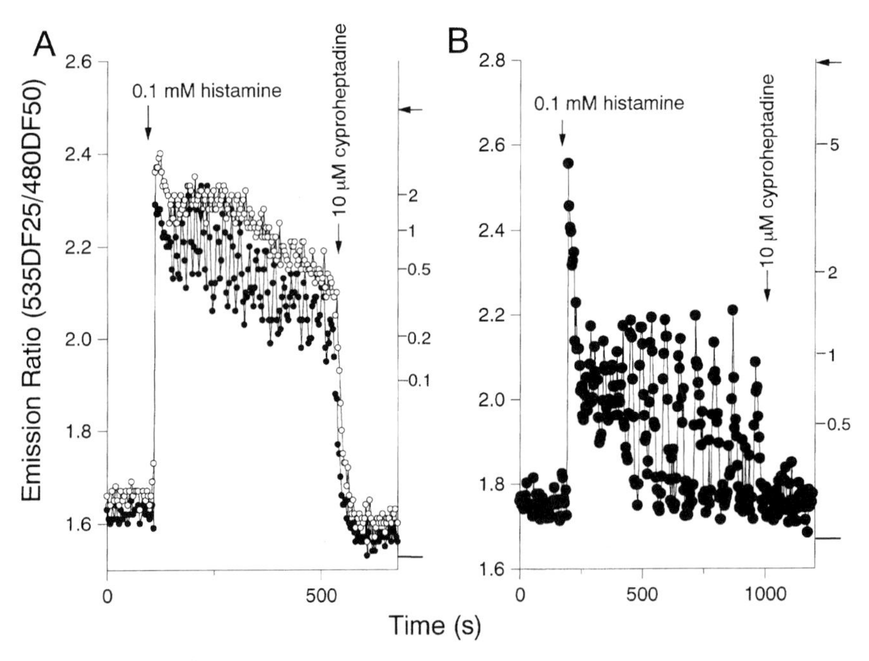

Figure 3 Typical Ca^{2+} transients reported by yellow cameleon-2.1 (A) and -3.1 (B) in HeLa cells induced by 0.1 mM histamine. A band pass filter (480DF50) was used for ECFP emission. The sampling interval was 4–5 sec. The arrows and horizontal bars on the right indicate R_{max} and R_{min} values, respectively. The right-hand ordinates calibrate $[Ca^{2+}]_i$ in μM. The concentrations of the indicators in the cells were 40–60 μM. Reproduced with permission from (9). Copyright National Academy of Sciences, USA.

between R_{min} and R_{max}, indicating that their amplitude was approximately equal to the K'_d of yellow cameleon-3.1 for Ca^{2+}, 1.5 μM. However, the relatively weak Ca^{2+} affinity also prevented detection of the sustained plateau elevation between spikes. Therefore the only apparent effect of cyproheptadine was to stop the oscillations. pH-related artefacts were not an issue in HeLa cells in these experiments, because agonist-induced $[Ca^{2+}]_i$ mobilization did not cause pH_i changes detectable by the pH indicator BCECF (data not shown). Correspondingly, comparisons of yellow cameleon-2 versus -2.1 and yellow cameleon-3 versus -3.1 showed no major differences in reported $[Ca^{2+}]_i$ attributable to the different pH sensitivities. On the other hand, yellow cameleon-2 or -3 in dissociated hippocampal neurons was perturbed by acidification following depolarization or glutamate stimulation. This problem was solved by the use of yellow cameleon-2.1 or -3.1 (9).

Protocol 1

Cytosolic Ca^{2+} imaging with yellow cameleons-2.1 and -3.1

Equipment and reagents

- Petri dish (35 mm diameter)
- Coverslip
- Inverted microscope (IX70, Olympus) with a cooled CCD camera (Micromax, Princeton Instruments)
- HeLa cells
- Lipofectin (Gibco BRL)
- cDNA
- Hanks' balanced salt solution (HBSS) containing 1.26 mM $CaCl_2$
- 1–5 μM of ionomycin
- Ca^{2+}-free medium

Method

1. Attach HeLa cells onto a coverslip in a Petri dish. Transfect cells in the dish with 1 μg of cDNA using Lipofectin.
2. Between 2–10 days after cDNA transfection, image HeLa cells on an inverted microscope (IX70) with a cooled CCD camera (Micromax). Expose cells to reagents at room temperature in HBSS containing 1.26 mM $CaCl_2$. Image acquisition and processing are controlled by a personal computer connected to the camera and a filter wheel (Lambda 10-2, Sutter Instruments, San Rafael, CA) using the program MetaFluor (Universal Imaging, West Chester, PA). An excitation filter wheel in front of the xenon lamp, and an emission filter wheel (Lambda 10–2) just below the CCD camera are also under computer-control. Although the excitation filter can be replaced by a fixed filter, the emission filter wheel (or another device to separate the two emission bandwidths) is required to image cameleons. Excitation light is from a 150 W xenon lamp passed through a 440DF20 (440 ± 10 nm) excitation filter. The light is reflected

Protocol 1 continued

onto the sample using a 455 nm long pass (455DRLP) dichroic mirror. The emitted light is collected with a x 40 (1.35 numerical aperture) objective, passed through a 480 ± 15 or 480 ± 25 nm band pass filter (480DF30 or 480DF50, donor channel) for ECFP, and a 535 ± 12.5 nm band pass filter (535DF25, FRET channel) for EYFP or EYFP-V68L/Q69K. Interference filters are from Omega Optical or Chroma Technologies (Brattleboro, VT). The band pass filters for excitation, and for donor and FRET channels are shown in Figure 2.

3 Define several factors for image acquisition. They are:

(a) Excitation power, which depends on the type of light source and neutral density filter.

(b) Numerical aperture of objective.

(c) Time of exposure to the light.

(d) Interval of image acquisition.

(e) Binning.

These last three factors should be considered in terms of which is pursued, temporal, or spatial resolution.

4 Choose intermediately-bright cells (see Protocol 3). Also the fluorescence should be uniformly distributed in the cytosolic compartment but excluded from the nucleus, as expected for a 74 kDa protein without targeting signals. Select regions of interest so that pixel intensities are spatially averaged.

5 At the end of an experiment, convert fluorescence signal into values of $[Ca^{2+}]$. R_{max} and R_{min} can be obtained in the following way. To saturate intracellular indicator with Ca^{2+}, increase extracellular $[Ca^{2+}]$ to 10–20 mM in the presence of 1–5 μM of ionomycin. Wait until fluorescence intensity reaches a plateau. Then to deplete the indicator of Ca^{2+}, wash the cells with Ca^{2+}-free medium (1 μM ionomycin, 1 mM EGTA, and 5 mM $MgCl_2$ in nominally Ca^{2+}-free HBSS). The in situ calibration for $[Ca^{2+}]$ uses the equation (10):

$$[Ca^{2+}] = K'_d[(R - R_{min}) / (R_{max} - R)]^{(1/n)}$$

where K'_d is the apparent dissociation constant corresponding to the Ca^{2+} concentration at which R is midway between R_{max} and R_{min}, and n is the Hill coefficient. The Ca^{2+} titration curve of yellow cameleon-2.1 could be fitted using a single K′d of 0.2 μM and a single Hill constant of 0.62. Therefore, use K'_d = 0.2 μM and n = 0.62 for yellow cameleon-2.1; K′d = 1.5 μM and n = 1.1 for yellow cameleon-3.1.

3.2 Concentration of yellow cameleons inside cells

Measurement of cameleon concentrations in cells is essential to quantify the trade-off between optical detectability and Ca^{2+} buffering. Here the methodology to separate the effects of path length and concentration is described.

Protocol 2

Preparation of recombinant yellow cameleon-3.1 protein in E. coli

Equipment and reagents

- French press
- Nickel-chelate columns (Qiagen)
- Centricon 30 (Amicon)
- T7 expression system $pRSET_B$/JM109(DE3) (Invitrogen)
- Isopropyl β-D-thiogalactoside

Method

1 Express the protein using the T7 expression system $pRSET_B$/JM109(DE3) (1). Grow cultures at room temperature, and induce protein expression by isopropyl β-D-thiogalactoside.

2 Lyse cells by a French press, and purify the polyhistidine-tagged proteins from the cleared lysates on nickel-chelate columns.

3 Concentrate the protein samples in the eluates by Centricon 30, so that the protein concentration is 5–50 μM.

Protocol 3

Estimation of concentration of indicators inside cells

Equipment and reagents

- Microchamber
- Coverslips
- Confocal microscope
- See *Protocol 2*

Method

1 Assemble three coverslips (150 μm thick) into a microchamber in which the thickness increased linearly from 0–150 μm. Its side-view is illustrated in *Figure 4*.

2 Fill the microchamber with the recombinant yellow cameleon-3.1 protein solution (see *Protocol 2*).

3 Image the microchamber on the microscope stage at the same instrument settings as those used for cells. The fluorescence intensities of ECFP (donor channel) and EYFP–V68L/Q69K (FRET channel) are proportional to the thickness of the sample. In a separate experiment using a confocal microscope, the thickness of HeLa cells can be estimated. For example, the perinuclear region, which gives brightest fluorescence using conventional microscopy is, in most cells, 5 μm thick. From the pre-stimulus intensities of ECFP and EYFP–V68L/Q69K of the perinuclear region in transfected HeLa cells to be imaged, the concentrations of yellow cameleon proteins inside the cells can be calculated, which usually range from < 5 μM to 1 mM. Usually intermediately-bright cells contain 50–200 μM cameleon proteins (see *Protocol 1*).

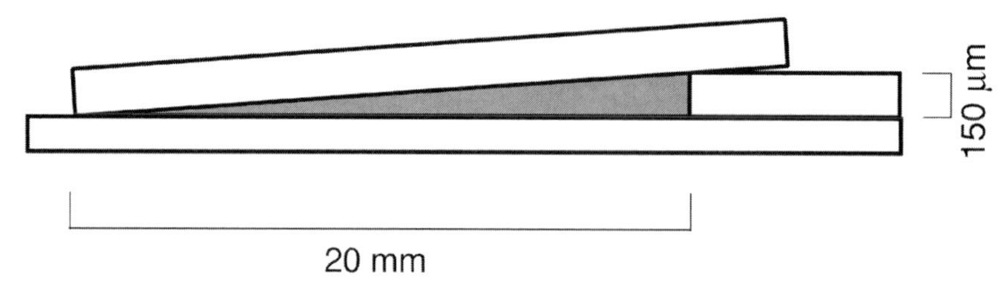

Figure 4 Microchamber containing defined concentrations and variable path lengths of indicators such as yellow cameleon-3.1 protein.

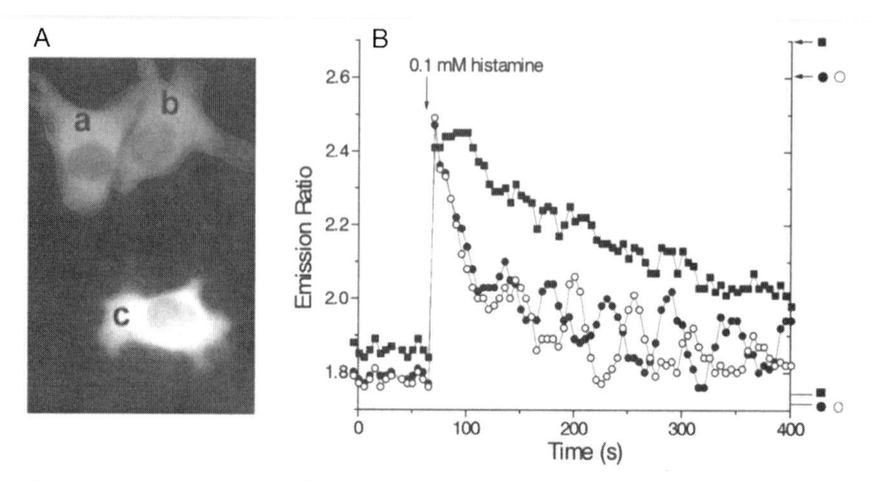

Figure 5 The effects of two different concentrations of yellow cameleon-3.1 on histamine-induced Ca^{2+} transients in HeLa cells. Indicator concentrations were estimated to be 150 μM in cells *a* and *b* and 500 μM in cell *c*. (A) Fluorescence images of the three cells. (B) Time courses of emission ratios (535DF25/480DF50) from cells *a* (filled circles), *b* (open circles), and *c* (filled squares). The arrows and horizontal bars on the right indicate R_{max} and R_{min} values, respectively. Reproduced with permission from (9). Copyright National Academy of Sciences, USA.

It is important to realize the ability of various concentrations of yellow cameleon-3.1 to buffer Ca^{2+} transients in HeLa cells. *Figure 5* compares $[Ca^{2+}]_i$ in three neighbouring cells expressing different levels of yellow cameleon-3.1 (*a* and *b*, 150 μM; *c*, 500 μM) during a 0.1 mM histamine challenge. Sharp $[Ca^{2+}]_i$ transients followed by $[Ca^{2+}]_i$ oscillations were observed in cells *a* and *b*. In contrast, $[Ca^{2+}]_i$ in cell *c* recovered slowly to the baseline over a period of several hundred seconds. Oscillations were never observed with $>$ 300 μM of yellow cameleon-3.1 in transfected HeLa cells. Cytosolic concentrations of cameleon below 20 μM are too dim to give good signal-to-noise ratios with our current instruments and could not be compensated by increasing the intensity of illumination because of YFP photochromism (Section 3.4).

3.3 Constitutive expression of cameleons and interaction with CaM and CaM-dependent enzymes

Stable HeLa and HEK293 cell lines constitutively expressing the improved yellow cameleons have been established. In those lines, the indicators are uniformly distributed in the cytosol but excluded from the nucleus, and $[Ca^{2+}]_i$ transients are similar to those in transiently transfected cells. The maintained exclusion from the nucleus and $[Ca^{2+}]_i$ sensitivity argue that the indicator proteins remain stable and unproteolysed, because cleavage between the CaM and M13 would generate split cameleon, whose components are small enough to enter the nucleus, and cleavage anywhere else would destroy $[Ca^{2+}]_i$-dependent FRET. That would add unwanted background signal on the donor and FRET channel.

CaM is involved in many basic cellular processes. If CaM is exogenously over-expressed as free form inside cells, many CaM-dependent pathways are altered, leading to a wide variety of physiological changes. For example, constitutive overexpression of CaM was shown to accelerate the rate of growth of cultured cells (10) and cardiomyocytes in transgenic mice (11). On the other hand, the fused CaM and M13 in yellow cameleons preferentially interact with each other upon Ca^{2+} elevation rather than with separate molecules of CaM or CaM-binding proteins. Therefore, if yellow cameleons are overexpressed, the primary effect is likely to be the unavoidable increase in Ca^{2+} buffering rather than specific perturbation of CaM-dependent signalling.

3.4 Photochromism of yellow fluorescent protein

If EYFP or EYFP–V68L/Q69K is excited too strongly, their fluorescence is reduced. This apparent bleaching is really photochromism (light-induced reversible isomerizations between two forms of a molecule having different absorption spectra), because the fluorescence recovers to some extent spontaneously and can be further restored by UV illumination (12). Intense excitation of yellow cameleons also causes such photochromism of the EYFP, which results in a decrease in the yellow:cyan emission ratio independently of Ca^{2+} changes. The extent of photochromism is determined by the excitation power, numerical aperture of objectives, and exposure time. It is necessary to optimize these factors for each cell sample to minimize the photochromism while still allowing a high signal-to-noise ratio. Because the photochromism is partially reversible, the sampling interval is another factor to be considered. Illumination at frequent intervals sometimes caused a decrease in the resting ratio values of yellow cameleon-3.1. To help the recovery of EYFP–V68L/Q69K, a pulse of broadband UV illumination (300 msec) could be inserted during the intervals between 440 nm excitations. This manoeuvre resets the resting emission ratio back to its original baseline, and enables us to follow histamine-induced oscillations in $[Ca^{2+}]_i$ without the decrease in the resting ratio values. But again the photodamage of the samples by UV illumination is a problem. A better solution is to bin pixels at the cost of spatial resolution. The increased signal-to-noise ratio permits decreasing the intensity of excitation light by a neutral density filter

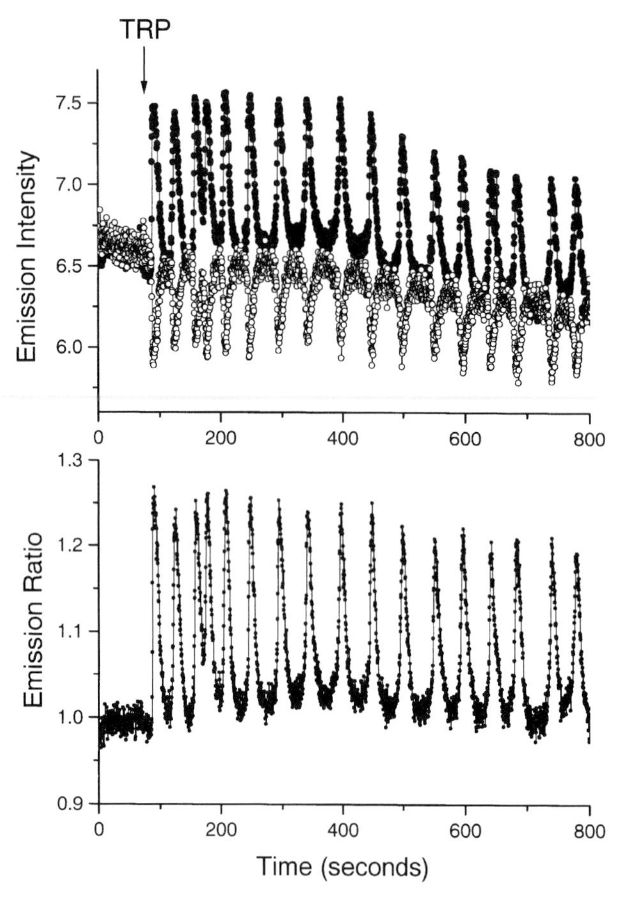

Figure 6 Dual wavelength photon counting measurement of $[Ca^{2+}]_c$ in a HEK293 cell which constitutively expresses yellow cameleon-2. Top two traces show the 480 nm (open circles) and 535 nm (closed circles) signals from yellow cameleon-2. Bottom trace is the emission ratio (535 nm/480 nm), representing $[Ca^{2+}]_c$. Data were collected at 3 Hz.

and observing $[Ca^{2+}]_i$ oscillations without significant photochromism of the indicators.

For dual emission ratio imaging of yellow cameleons, consecutive data gathering has been described so far in this chapter; images are created by alternating filters in the emission path. A photometric system using photomultiplier tubes permits measurement of photons on fast time scales, because it is not restricted by the frame rate of cameras. Also because less excitation light is needed to obtain sufficient signal-to-noise ratio, the system permits very long measurements. Its drawback is that it provides only temporal information with no spatial dimensions. *Figure 6* shows dual wavelength photon counting measurements of $[Ca^{2+}]_c$ in a HEK293 cell stably expressing yellow cameleon-2. In this experiment, at = 90 s, thrombin receptor activating peptide (TRP) was added at a final concentration of 30 μM. In this particular cell, Ca^{2+} oscillations were measured continuously for over five hours.

Protocol 4

Simultaneous measurements of both wavelengths of yellow cameleons

Equipment and reagents

- Coverslip
- Zeiss inverted microscope (Axiovert 135)
- 75 W xenon arc lamp, coupled to a monochromator (PTI)
- PTI microscope photometer system (D104)
- HEK293 stable-transformant cells

Method

1 Grow the HEK293 stable-transformant cells on coverslips.

2 Mount the coverslip on a Zeiss inverted microscope equipped with a x 63, 0.9 NA objective.

3 Excitation (425 ± 5 nm) is by means of a 75 W xenon arc lamp, coupled to a monochromator (PTI). The excitation light is coupled to the epifluorescence port of the Axiovert using an optical fibre, and reflected onto the cell using a 70/30 neutral beam splitter.

4 Emission is detected at the camera sideport. This port is equipped with a PTI microscope photometer system (D104) that allows selection of a single cell by masking of other parts of the image. The resulting emission is split by a dichroic mirror at 510 nm and passed through 480 ± 15 nm and 535 ± 15 nm filters to yield the cyan and yellow fluorescence signals, respectively.

5 Detection is by means of two PTI type 810 photon counting PMTs, interfaced to a PC via a Labmaster I/O board. Data are gathered and the ratio is calculated on-line using FELIX software (PTI) at three samples per second.

4 Measurements of nuclear Ca^{2+} concentration ($[Ca^{2+}]_n$)

4.1 $[Ca^{2+}]_c$ versus $[Ca^{2+}]_n$

It has been controversial whether nuclear Ca^{2+} concentration ($[Ca^{2+}]_n$) is regulated independently of the cytosolic Ca^{2+} concentration ($[Ca^{2+}]_c$) (13). Comparative measurements of $[Ca^{2+}]_n$ and $[Ca^{2+}]_c$ have been performed using synthetic fluorescent chelators (13–15). However, the cytosolic signal is often contaminated by that from intracellular organelles such as ER, into which the chelators can be compartmentalized. The Ca^{2+}-sensitive photoprotein, aequorin has been localized specifically in nucleus and cytosol (16–19), but it has proved difficult to image $[Ca^{2+}]_n$ and $[Ca^{2+}]_c$ in individual cells.

4.2 Targeting yellow cameleon-3.1 to nucleus

Addition of a nuclear localization signal to yellow cameleon-3.1 yields a Ca^{2+} indicator, yellow cameleon-3.1nu, the fluorescence of which is tightly localized to nuclei, but excluded from nucleoli. Because exogenously-expressed proteins accumulate in the nucleus at a higher concentration than in the cytosol, usual transfection procedures give a very high level of expression of the indicator protein in the nucleus. When yellow cameleon-3.1 and yellow cameleon-3.1nu are co-expressed in HeLa cells, it is necessary to control the amounts of the two cDNAs for co-transfection. To balance the indicator's concentration between the two compartments, the cDNAs of yellow cameleon-3.1 and yellow cameleon-3.1nu should be used in the ratio 10 : 1. Then it is possible to compare the two Ca^{2+} concentrations without concern over difference in Ca^{2+} buffering.

4.3 Sectional images of $[Ca^{2+}]_c$ and $[Ca^{2+}]_n$ by two-photon excitation

Since UV laser can excite blue cameleons that contain EBFPs as the donors, it is possible to see confocal $[Ca^{2+}]_i$ images by conventional confocal microscopy. There is a one-photon laser line (He/Cd, 442 nm) for the efficient and selective excitation of ECFP rendering conventional confocal microscope of use for imaging yellow cameleons. However, the use of multiphoton optical absorption is more attractive to mediate excitation in laser scanning microscopy (20). It has been found that two-photon excitation with 780–800 nm pulses selectively excites ECFP in preference to EYFP (21). Two-photon excitation microscopy allows us to get optically-sectioned images of $[Ca^{2+}]_c$ and $[Ca^{2+}]_n$ in individual cells, which express both yellow cameleon-3.1 and yellow cameleon-3.1nu. To date, no obvious concentration gradient of Ca^{2+} across the nuclear membrane of HeLa cells has been evident in the images in high time-resolution video sequences (21).

5 Measurements of free Ca^{2+} concentration inside the endoplasmic reticulum ($[Ca^{2+}]_{er}$)

Measurements of $[Ca^{2+}]_{er}$ using small synthetic chelators are subject to a considerable number of uncertainties in dye localization, Mg^{2+} interference, and intactness of the perfused cytosolic compartment. Measurements using ER-targeted aequorin can only be done on cell populations and are complicated by the opposing abilities of Ca^{2+} both to stimulate light output and to irreversibly destroy the indicator. Yellow cameleon-3er and -4er were engineered to reside in the lumen of endoplasmic reticulum (ER) by addition of a signal sequence at the N-terminus and a KDEL signal for ER retention at the C-terminus (see *Figure 1*). These indicators permit measurements of agonist-induced changes in the free Ca^{2+} inside the ER in individual, intact, non-perfused cells.

Figure 7 shows $[Ca^{2+}]_{er}$ reported by yellow cameleon-3er in two neighbouring HeLa cells. Because of the intermediate affinity for Ca^{2+} (K'_d, 1–2 μM), yellow cameleon-3er was in one cell saturated ($[Ca^{2+}]_{er} > 100$ μM) at resting state; the

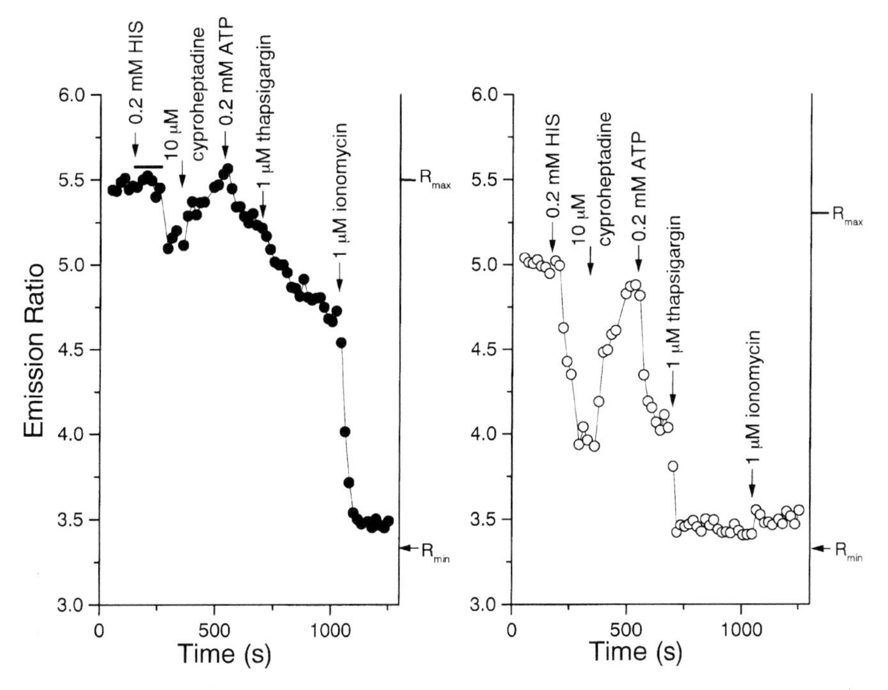

Figure 7 The 535 to 480 nm emission ratios from yellow cameleon-3er in two HeLa cells. The right-hand ordinates give R_{min} (arrow) and R_{max} (bar) values.

indicator can not measure $[Ca^{2+}]_{er}$ over 100 μM. The initial emission ratio value matched R_{max}, and there was a time delay in response to the Ca^{2+}-mobilizing effect of histamine (*Figure 7*, *left*, the delay is indicated by a horizontal line). In the other cell, basal $[Ca^{2+}]_{er}$ was lower (less than R_{max}). Therefore, resting $[Ca^{2+}]_{er}$ was about 60 μM, and there was much less time delay in response to histamine (*Figure 7*, *right*). Yellow cameleon-3er can report the depletion of $[Ca^{2+}]_{er}$ well.

To measure resting $[Ca^{2+}]_{er}$ values above 100 μM, the lowest-affinity indicator, yellow cameleon-4er should be used. It has a K'_d value of around 500 μM, but can not measure $[Ca^{2+}]_{er}$ below 100 μM. With this indicator, the emission ratio starts well below R_{max}, and falls immediately upon histamine stimulation. However, once $[Ca^{2+}]_{er}$ is depleted below 100 μM, the response to antagonist application is delayed (data not shown).

In conclusion, to monitor the resting and depleted $[Ca^{2+}]_{er}$, it is currently necessary to combine results from the two indicators. Construction of an ideal indicator for $[Ca^{2+}]_{er}$ measurement which reports a wider range of Ca^{2+} concentration with a K'_d value of about 100 μM is underway.

References

1. Miyawaki, A., Llopis, J., Heim, R., McCaffery, J. M., Adams, J. A., Ikura, M., *et al.* (1997). *Nature*, **388**, 882.
2. Grynkiewicz, G., Poenie, M., and Tsien, R. Y. (1985). *J. Biol. Chem.*, **260**, 3440.

3. Brini, M., Marsault, R., Bastianutto, C., Alvarez, J., Pozzan, T., and Rizzuto, R. (1995). *J. Biol. Chem.*, **270**, 9896.
4. Porumb, T., Yau, P., Harvey, T. S., and Ikura, M. (1994). *Protein Eng.*, **7**, 109.
5. Ikura, M., Clore, G. M., Gronenborn, A. M., Zhu, G., Klee, C. B., and Bax, A. (1992). *Science*, **256**, 632.
6. Zolotukhin, S., Potter, M., Hauswirth, W., Guy, J., and Muzyczka, N. (1996). *J. Virol.*, **70**, 4646.
7. Gao, Z. H., Krebs, J., VanBerkum, M. F. A., Tang, W. J., Maune, J. F., Means, A. R., *et al.* (1993). *J. Biol. Chem.*, **268**, 20096.
8. Llopis, J., McCaffery, J. M., Miyawaki, A., Farquhar, M. G., and Tsien, R. Y. (1998). *Proc. Natl. Acad. Sci. USA*, **95**, 6803.
9. Miyawaki, A., Griesbeck, O., Heim, R., and Tsien, R. Y. (1999). *Proc. Natl. Acad. Sci. USA*, **96**, 2135.
10. Ramussen, C. D. and Means, A. R. (1987). *EMBO J.*, **6**, 3961.
11. Gruver, C. L., Francesco, D., Goldstein, M. A., and Means, A. R. (1993). *Endocrinology*, **133**, 376.
12. Dickson, R. M., Cubitt, A. B., Tsien, R. Y., and Moerner, W. E. (1997). *Nature*, **388**, 355.
13. Malviya, A. N. and Rogue, P. J. (1998). *Cell*, **92**, 17.
14. Bolsover, S. R., Silver, R. A., and Whitaker, M. (1993). In *Electronic light microscopy* (ed. D. Shotton), pp. 181–210. Wiley-Liss, New York.
15. Brown, G. R., Koehler, M., and Berggren, P. (1997). *Biochem. J.*, **325**, 771.
16. Brini, M., Murgia, M., Pasti, L., Picard, D., Pozzan, T., and Rizzuto, R. (1993). *EMBO J.*, **12**, 4813.
17. Brini, M., Marsault, R., Bastianutto, C., Pozzan, T., and Rizzuto, R. (1994). *Cell Calcium*, **16**, 259.
18. Badminton, M. N., Campbell, A. K., and Rembold, C. M. (1996). *J. Biol. Chem.*, **271**, 31210.
19. Badminton, M. N., Kendall, J. M., Rembold, C. M., and Campbell, A. K. (1998). *Cell Calcium*, **23**, 79.
20. Denk, W. and Svoboda, K. (1997). *Neuron*, **18**, 351.
21. Fan, G. Y., Fujisaki, H., Miyawaki, A., Tsay, R.-K., Tsien, R. Y., and Ellisman, M. H. (1999). *Biophys. J.*, **76**, 2412.

Chapter 2

Photometry, video imaging, confocal and multi-photon microscopy approaches in calcium signalling studies

Peter Lipp, Martin D. Bootman, and Tony Collins
Laboratory of Molecular Signalling, The Babraham Institute, Babraham Hall, Babraham, Cambridge CB2 4AT, UK.

1 General introduction

Calcium (Ca^{2+}) is one of the most ubiquitous intracellular second messengers in living cells, and controls processes as diverse as fertilization, secretion, and muscle contraction (for review see ref. 1). The reason why a simple ion such as Ca^{2+} can regulate a vast array of cellular activities is that it is a very versatile messenger. A multitude of Ca^{2+} signals, varying in kinetics, amplitude, frequency, and spatial dimension, are utilized by different types of cells. This variability of intracellular Ca^{2+} signals means that there is no single technique that can be used to study Ca^{2+} in all situations. Instead, there are many types of apparatus, each with advantages and drawbacks that allow researchers to visualize Ca^{2+} changes in their particular cellular system. This chapter describes some of the most common approaches to monitoring or manipulating cellular Ca^{2+} changes using fluorescence, and attempts to give hints for designing experiments, trouble-shooting practical problems, and obtaining real signals. Since many of the protocols and problems associated with photometry and video imaging are the same, these two techniques are dealt with together. Confocal and multi-photon applications are discussed separately. The recent development of chameleons as Ca^{2+} detectors using fluorescence resonance energy transfer (FRET) are not described here. For a detailed account of the use of FRET to monitor Ca^{2+}, please see Chapter 1.

2 Fluorescent Ca^{2+} indicators

Fluorescent Ca^{2+} indicators are molecules whose optical properties change when they bind Ca^{2+}. Usually either the excitation or the emission spectrum (or both) are changed upon Ca^{2+} binding. Several different strategies can be employed to introduce indicators into cells. Indicators can be classified as ratiometric or non-ratiometric, depending on whether Ca^{2+} binding changes only the magnitude of their absorption or emission, or whether their spectra shift horizontally along the wavelength axis (i.e. they change colour). This section discusses protocols for loading Ca^{2+} indicators and the factors that determine the choice of indicator. A short section on preparing adherent cells for imaging is also included.

2.1 Loading Ca^{2+} indicators

The plasma membrane is the principal barrier to fluorescent indicators, and various techniques have been developed to allow the incorporation of the indicators into the cytoplasm whilst maintaining cell integrity. In their Ca^{2+}-sensitive forms, the Ca^{2+} indicators are all hydrophilic charged (usually $-COO^-$ groups) compounds that cannot readily cross lipid membranes. Incubating cells with versions of indicators where the charged groups have been masked by esterification is the most commonly used loading technique. Since they are highly lipophilic, stock solutions of the indicators esters should be made using DMSO (*Protocol 1*). Including a detergent, such as Pluronic, in the stock solution can very often help loading with esters since it prevents formation of indicator micelles. Once the esters are in the cytoplasm, they are cleaved by endogenous esterases and released in their Ca^{2+}-sensitive forms.

Protocol 1

Making up a stock solution of esterified Ca^{2+} indicators

Equipment and reagents

- Pipette tip or sonicator
- Pluronic F-68
- DMSO
- Solid indicator

Method

1. Dissolve 20% (w/w) Pluronic F-68 in an appropriate volume of dry DMSO.
2. Stir this mixture until the Pluronic is dissolved while simultaneously gently warming the solution. The temperature should not exceed 40 °C.
3. Add an appropriate volume of the DMSO/Pluronic mixture to the indicator solid. A useful final concentration of the stock solution is 1 mM.
4. Make sure that the entire indicator has been dissolved in the DMSO/Pluronic mixture by agitating the solution through a pipette tip or sonication.

Although convenient, the use of esters to load cells with indicators is not without problems. In particular, it is easy to overload cells with indicators, thus increasing the intracellular buffering and damping Ca^{2+} signals. In addition, ester loading can suffer from the drawback that some of the indicator becomes compartmentalized into organelles such as mitochondria and endoplasmic reticulum (see Chapter 6). Such compartmentalization will complicate calibration of Ca^{2+} signals. Furthermore, many cells have the capacity to rapidly clear indicator molecules from their cytoplasm. Another problem is that of incomplete ester hydrolysis. The presence of the ester groups prevents the indicator from chelating Ca^{2+} and makes it Ca^{2+} insensitive. A Ca^{2+}-insensitive, but still fluorescent indicator will contribute to the overall signal, resulting in an underestimation of Ca^{2+} concentration. Incomplete ester hydrolysis occurs in cells that have low esterase activity. This can often be overcome by extending the de-esterification period.

Minimal indicator loadings are preferable to reduce effect on cellular buffering. Performing ester loadings at 20–22 °C, before adjusting the cells back to the desired experimental temperature, can usually reduce compartmentalization and clearance of indicator from cells. The use of sulfinpyrazone (100 μM), probenicid (1 mM), or leakage-resistant indicators (e.g. fura-2/PE3; TefLabs) can prevent indicator loss from most cell types. *Protocol 2* describes our typical loading procedure for esterified forms of indicators that applies for most excitable and non-excitable cells. The most appropriate incubation times and concentrations may vary between indicators and cell types, and should be worked through empirically.

Protocol 2

Loading of cells with the ester form of Ca^{2+} indicators

Equipment and reagents

- Coverslip/cuvette
- See Protocol 1
- Extracellular solution
- Cells

Method

1. Make up a 2 μM solution of the indicator by adding 4 μl of the 1 mM stock (see Protocol 1) to 2 ml of an appropriate extracellular solution. Make sure the indicator stock is completely dissolved.
2. Since constituents of serum can bind to indicators and reduce their free concentration, cells should be washed free of culture medium prior to loading.
3. Put the 2 ml of the indicator solution (or whatever volume is required to fill the coverslip/cuvette bearing the cells) and incubate for 30 min, preferably at room temperature (i.e. 20–22 °C) and in a dark place.
4. Remove the indicator solution, but keep it for additional loadings (can usually be reused three to five times, depending on the actual cell density). Replace with indicator-free extracellular solution and incubate the cells for an additional 20–30 min at room temperature in the dark to allow complete de-esterification of the indicator.

A variety of alternative methods for loading cells with the free acid (Ca^{2+}-sensitive) forms of indicators have been employed, including microinjection, infusion from whole-cell patch pipettes, pinocytosis, and transient plasma membrane permeabilization using osmotic shock or electroporation. The suitability of these different techniques is often dependent on the cell type. The major advantages of loading free acid forms of the indicators is that the intracellular Ca^{2+} concentration can be more precisely controlled and the indicator is often restricted to the cytoplasmic compartment.

It should be noted that even the free acid form of the indicators can be sequestered into cellular organelles. This problem can be largely circumvented by injection of dextran-conjugated indicators. Dextran molecules are inert sugars of varying sizes, typically 10 kDa to 70 kDa. However, it has been found that the conjugation of the indicator molecules to the dextran sugar backbone is random, and the affinity of the individual indicator molecules for Ca^{2+} varies depending upon their position on the dextran molecule. Thus the overall affinity of a dextran-conjugated Ca^{2+} indicator may vary from batch to batch and needs to be determined experimentally if very accurate values of Ca^{2+} are required.

2.2 Choice of Ca^{2+} indicator

2.2.1 Excitation and emission wavelengths

A wide variety of Ca^{2+} indicators are currently available, with excitation and emission spectra ranging from UV to visible colours. In many situations, however, the choice of indicator is limited by experimental or hardware considerations. For example, confocal microscopy generally uses lasers to provide sufficient illumination. Although the light output can be intense, lasers provide only a limited number of wavelengths of excitation light. In contrast, photometry and video imaging tend to use so-called 'white-light' sources, which emit a broad spectrum, and thus allow a wider choice of indicators.

2.2.2 Single- versus dual-wavelength indicators

Ca^{2+}-sensitive fluorescent indicators can be broadly divided into single- and dual-wavelength indicators on the basis of their spectral changes in response to Ca^{2+} elevation. Single-wavelength Ca^{2+}-sensitive indicators (e.g. fluo-3) change their emission intensity upon binding Ca^{2+}. However, the intensity of emission is also proportional to the indicator concentration, so careful calibration is required with these indicators. Bleaching of the indicator over the period of the experiment will change the effective indicator concentration and can complicate calibration of Ca^{2+} responses.

With dual-wavelength indicators, problems of uneven distribution and changes in indicator concentration are circumvented. When Ca^{2+} binds to a dual-wavelength indicator molecule, the excitation or emission spectrum changes. For example, with Indo-1, the Ca^{2+}-free molecule has a peak emission at 490 nm, whilst the Ca^{2+} bound molecule emits optimally at 405 nm. By ratioing the emission of Indo-1 at these two wavelengths, a measure of Ca^{2+} can be made that is independent of the indicator concentration. Dual-wavelength indicators also display an excitation or emission wavelength where the indicator is insensitive to Ca^{2+}. This is known as the 'isosbestic point', and can be used to measure the indicator concentration or to monitor interaction of the indicators with other metal ions, such as Mn^{2+}.

Two single-wavelength indicators (e.g. fura red and fluo-3) can be combined to give pseudo-dual wavelength indicator (2). This is not as satisfactory as a true dual-wavelength indicator since several assumptions are made: both indicators need to be similarly distributed in the cytosol, and they must be bleached or extruded from the cell at similar rates. In addition, the affinities of the indicators for Ca^{2+} must be similar. Some of these problems are solved when using a single Ca^{2+} indicator (e.g. fluo-3 or Calcium Green) together with a Ca^{2+}-insensitive reference fluorescence (3). Another attempt to overcome the lack of a visible ratiometric indicator is to employ a Ca^{2+}-sensitive indicator that has been conjugated to a dextran molecule along with a reference dye such as Texas Red. By linking the indicators in this way they will obviously have the same intracellular distribution and leak rates, however, bleach rates may still differ and the conjugation to a large dextran necessitates a delivery method other than ester-loading.

2.2.3 Other considerations

The actual choice of indicator is determined by several factors; whether a dual- or single-wavelength indicator is necessary, the affinity of the indicator, and its dynamic range. Dual-wavelength indicators are generally preferable in that the ratio output (see above) is less prone to problems associated with indicator distribution, bleaching, and leakage. However, when very rapid image acquisition is required, the time for image pairs to be taken may be unacceptable, so that a single-wavelength indicator is most suitable. In addition, Ca^{2+}-sensitive dual-wavelength indicators are only available with UV excitation. The use of UV light has introduces problems such as enhanced cellular autofluorescence and cell damage, which are greatly reduced when using single-wavelength indicators that are excited by visible light.

Another consideration for the choice of indicator is that of the dissociation constant (K_d) of the indicator. The K_d of an indicator reflects the indicators affinity for Ca^{2+}. The higher the K_d, the lower the affinity of the indicator for Ca^{2+}. Typically an indicator is only sensitive to changes of Ca^{2+} one order of magnitude above and below its K_d. Concentrations of Ca^{2+} in the cell can range from tens of nanomolar to several micromolar, this means that the indicator must be matched to the magnitude of Ca^{2+} changes of interest.

The dynamic range of an indicator can be thought of as how much the fluorescence intensity changes upon binding of Ca^{2+}. The greater the dynamic range, the higher the signal-to-noise ratio and the higher the chance to pick up small changes of Ca^{2+}. In our hands, fluo-3 offers the best dynamic range of the currently available indicators, and allows both small amplitude elementary Ca^{2+} signals (see below) and larger global Ca^{2+} waves to be monitored equally well.

2.3 Using adherent cells

Although fluorescent indicators can be used with adherent or freely-floating cells, in many situations cellular physiology is best preserved if the cells are adhered to glass coverslips before use. This procedure causes many cells to flatten, which is a beneficial situation for imaging and also allows perfusion without cell movement. Our laboratory generally uses 40 mm diameter glass coverslips (BDH) with ~ 170 μm thickness (a suitable thickness for most modern microscopes), which can be sterilized by simply washing in ethanol and phosphate-buffered saline. A large diameter coverslip is preferable as it gives a wider area to search for suitable cells, and allows perfusion and suction lines to be remotely located. Most cultured cell lines will adhere straight on to the glass. However, some cell types, and especially primary cultures, may require a substratum such as poly-L-lysine. Plant cells can be trickier to deal with, and the growth conditions for these cell types will depend very much on the particular tissue. Germ tubes of fungal spores are often adherent themselves, as they are producing an adhesive extracellular matrix and only need to be allowed to germinate. Some fungal species prefer a hydrophobic surface so silanization of coverslips may be required. Some pollen tubes require growth in a low density agar medium to allow them to be viewed, whilst leaf epidermal strips may simply be mounted on a coverslip.

3 Photometry and video imaging of Ca^{2+}

Both photometry and video imaging quantify the fluorescence signal from a specimen loaded with a Ca^{2+}-sensitive indicator. A good portion of the equipment, i.e. microscope, illumination source, is the same for both techniques. The main difference between photometry and video imaging is the detector used to quantify the emitted fluorescence. In photometry the detector is a photomultiplier tube (PMT), whilst in video imaging the detector is a camera. This difference determines both the cost difference between the two systems (cameras are far more expensive than PMTs), and also the type of data generated. Photometry can provide data with a high temporal resolution, but no direct spatial information of the underlying cellular Ca^{2+} responses. With most video imaging systems, a medium level of spatial and temporal resolution can be achieved (confocal microscopy can be thought of as providing high spatial resolution with low/high temporal resolution—see below). This section discusses the use of PMT- and camera-based systems for monitoring Ca^{2+} signals.

3.1 Photomultiplier tubes

Photomultipliers (PMT) convert single-photon events into an electrical current. The photons emitted from the specimen are directed to the front of the PMT where they hit a photocathode, which subsequently emits electrons from its back face. The characteristics of the photocathode determine the quantum efficiency (usually about 20–30% of incident photons result in the emission of an electron). PMTs often have a spectral bias determining whether blue or red photons are more efficiently detected. The current arising from PMTs can be quantified in two different ways. Most commonly, the current is converted to a voltage that is measured by an analogue-to-digital converter card attached to a host computer. Alternatively, at low light intensities, each photon event generates a discrete pulse of current which can be individually counted. This 'photon counting mode' requires sophisticated circuitry, and can only be done if the photons arrive more slowly than the duration of the current pulse generated by a single-photon event (typically tens of nanoseconds). This situation does not typically arise when using bright fluorescent indicators and is more often used with chemiluminesence measurements.

3.2 Cameras

Most modern cameras are based on charged coupled device (CCD) technology, although some older type 'tube' cameras are still available. The CCD can be thought of as a grid of light-sensitive areas—'diodes'. As an image from the specimen forms on the CCD, varying numbers of photons fall on each diode from the dark or light parts of the image. An electrical charge is generated at each diode proportional to the number of incident photons, which is used to generate the output signal and subsequent image. Signal amplifiers ('intensifiers') located before the CCD chip can enhance the light sensitivity of cameras, but this will generally reduce the spatial resolution and increase noise in the image.

3.2.1 Analogue cameras

Analogue cameras output a standard analogue video signal, like the one from home video recorders. This video standard has a defined vertical resolution and image frequency. In Europe the standard monochrome signal (CCIR format) has a vertical resolution of 625 lines and images are generated at the rate of 25 per second. In North America and Japan the format (RS-170 format) produces 525 line images at 30 per second. Video cameras are bound by these formats, which limits camera performance. However, the standard output signal does mean that any analogue video camera can be used with any analogue video 'frame grabber' (a device that converts the analogue video signal into digital data), and any software that supports such frame grabber.

3.2.2 Digital cameras

Digital cameras are not tied to a video standard and are therefore more versatile in their output. The signal from digital cameras is often in RS422 format and requires a suitable computer interface card, although both SCSI- and RS232C (serial port)-interfaced cameras are available. The rate of output from the camera can be varied; a slow output allows longer integration time (i.e. a longer period when the CCD chip is exposed to the incident light) resulting in a better signal-to-noise ratio.

With many digital cameras it is possible to monitor an area of interest within the full image. By excluding unnecessary areas of the CCD chip in this way camera read-out times can be increased, thereby increasing image acquisition times. Another useful property of digital cameras is that they allow 'binning', where a selectable number of neighbouring diodes are effectively combined into a single diode. This increases the surface area of each diode, thus enhancing light sensitivity and allowing slower integration times and higher read-out speeds. Obviously, binning reduces spatial resolution. When a digital camera bins all its diodes, then video imaging becomes photometry!

3.3 Data format—counts and images

The data generated by photometry is usually a voltage or photon counts per unit of time, and is displayed as a trace. The data generated by video imaging is a sequence of digital images. A digital image is formed as a two-dimensional array of picture elements (pixels). The number of pixels that make up an image is dependent on the CCD chip for digital cameras or the frame grabber for analogue cameras (typically 768 × 512). The cameras used in video imaging measure intensities of fluorescence. Therefore monochrome images are generated, and no colour information is recorded other than that inferred from use of an emission filter.

3.4 Intensity depth

If one thinks of a grey spectrum from black to white, it is black at one end, white at the other with an infinite number of greys in between. For imaging this infinite analogue spectrum needs to be digitized, i.e. split into chunks of equal intensity difference. Such chunks are referred to as grey levels. A 1-bit (i.e. either 0 or 1) image can represent two grey levels, and divides the grey spectrum into a digital black and white image. If the number of 'bits', i.e. the 'bit-depth', is increased there is a corresponding increase in the number of greys levels. Since the human eye can discriminate less than 50 shades of grey, an 8-bit image gives the appearance of a continuous intensity profile. The camera and its settings will determine the bit-depth of an image. The most common camera formats are 8-bit (256 grey levels), 12-bit (4096 grey levels), and 16-bit (65 536 grey levels). Many modern digital cameras are capable of providing 12-bit and 16-bit resolution allowing greater discrimination between light intensities. However, this extra resolution increases the amount of memory needed to store each image, and is not necessary for many applications.

3.5 Look-up tables

To more clearly display changes in the intensity of fluorescent indicators, the grey scale images from monochrome sources are commonly colour-coded using a 'look-up table'. Put simply, each grey value is assigned a colour on the basis of how much red, green, and blue the pixel should contain. Look-up tables are often simply text files which contain four data columns. The first represents the grey scale values. The other three columns contain data indicating the corresponding intensity for red, green, and blue. Colouring grey scale images in this way can help the observer discriminate changes in Ca^{2+}. The usual convention is that cold colours (black, blue, and green) encode low Ca^{2+} concentrations, whilst warmer colours (yellow, red, and white) denote high Ca^{2+} concentrations.

3.6 Photometry versus video imaging

The major advantage of photometry over video imaging is the speed of data acquisition. Higher sampling speeds are obtainable with PMTs compared to video imaging because there is only one measurement to be made per moment in time compared to the 393 216 measurements in a 768 × 512 image. Of course, the fast sampling rate of photometry is at the expense of spatial information, since the whole field of view is averaged. With photometry it is often necessary to have either only a single cell in the field of view, or to use an iris diaphragm to exclude unwanted areas, so that the data is not complicated by asynchronous responses from multiple cells. The results of photometry experiments can be relatively simply calibrated. Calibration of fluorescent indicators for photometry measurements is described in Section 10.1 and *Protocol 5* of this chapter. Photometry also allows *simultaneous* dual-emission ratio experiments (e.g. Indo-1, SNARF) with the relatively modest expense of a second PMT, compared to a second camera. The most significant advantage video imaging has over photometry is providing spatial information on the change in indicator intensity within a single cell. Furthermore, multiple cells can be viewed simultaneously, thus increasing the sample number for subsequent statistics.

3.7 Brightness and gain settings (see also Section 6.1.3)

Aside from the genuine signal from the indicator in the specimen, there will also be a background signal. This signal should be quantified and subsequently subtracted from the raw data. The background signal comprises random noise from the detector and also fluorescence signal from leaked indicator and stray light. A measure of the background can be obtained from an illuminated field of view with no cells present assuming that the cellular autofluorescence is negligible. Background information must be obtained for both wavelengths if using a dual-wavelength indicator. Cells with strong autofluorescence (in particular plant and fungal cells) provide extra problems with regard to background signal and may not be corrected for in the final image without sophisticated image processing. It is necessary to examine cells in the absence of indicators to determine whether autofluorescence will be a problem for the desired imaging conditions.

Part of the background signal comes from intrinsic noise in the detector. A measure of the detector noise, 'dark current', can be obtained simply by shutting off the excitation illumination. Some cameras have an adjustable 'black level' (brightness or off-set) which can be modulated to off-set the dark current. The black level sets the intensity range at which the camera starts to detect. Ideally, the black level should be modulated so that there is a barely detectable level of noise in the absence of excitation.

Black level correction is not a feature of all cameras. However, since the background signal can be calculated and subtracted this is not a major problem. Being able to adjust the black level does increases the grey levels available for imaging though. For example, if the background signal is say 25, there are only 230 grey levels left for the image. Black level controls are also found on confocal microscopes and perform the same function (see Section 6.1.3).

The gain (or contrast setting) of a camera/PMT adjusts how much the signal is amplified. However, a higher gain will give a noisier final image. A balance must be achieved so that there is adequate signal with acceptable noise. The noise in the camera or PMT, i.e. the dark current, arises from several sources, but is primarily due to thermal emission of electrons from a photocathode in a PMT or charge build-up on a CCD. The dark current normally only contributes a small percentage to the charge on the CCD or current from a PMT. However, noise becomes an increasing percentage of the true signal as the signal weakens (the noise doesn't actually increase—it just becomes a more significant part of the overall signal). Increasing the gain does not increase the noise at source, i.e. there are no extra electrons generated at the photocathode or charges built up in the CCD, but increasing the gain does amplify the noise along with the signal making them both more visible. The black level or off-set of the camera/PMT must be adjusted after changing the gain. The gain control is important in preventing saturation of the detector (i.e. a signal that is too bright). If the camera/ PMT saturates, the gain (or alternatively the excitation illumination) must be reduced.

4 Troubleshooting photometry and video imaging

The following sections provide hints on how to adjust hardware parameters in order to improve fluorescence detection.

4.1 Image too dark

4.1.1 Objective

Normally, the objective of choice should be a high numerical aperture (NA) plan apochromatic objective (i.e. corrected for both chromatic and spherical aberration). However, such correction is at the expense of transmission, especially at shorter wavelengths (such as that of fura-2, 340 nm). 'Fluor' objectives are designed to give high transmission at all wavelengths. If the signal is too low, it may be better to accept the lack of spherical and chromatic correction of a fluor objective (of equal high NA) in order to obtain a good signal.

4.1.2 Filter set

By increasing the bandwidth of the excitation or emission filter, a brighter image will usually be obtained. To minimize indicator bleaching it is better to maximize the collection of emitted light by widening the bandwidth of the emission filter rather than the excitation bandwidth. Long pass (LP) filters (those that pass light longer than the designated wavelength) are generally preferable as emission filters. With dual-emission indicators however, band pass (BP) filters must be used. These pass light of the designated wavelength plus and minus half the second number in the designation, e.g. a 490BP10 filter passes light of wavelengths 485–495 nm. (Note: This is not *exactly* true, but is effectively true for a good (i.e. expensive) filter. A graph of intensity passed through a filter versus wavelength (transmission profile) of light appears as somewhere between a square wave and normal distribution, where in this range an individual filter's transmission profile occurs depends upon the quality of the filter, the square wave being preferable. The specification of a filter (e.g. 490BP10) designates the optimally transmitted wavelength (i.e. 490 nm) and the range of wavelengths transmitted at half the intensity of the optimum transmission, the full width, half maximum (FWHM). It is advisable to use a spectrometer to determine the optical properties of your filters.)

It is often not desirable to greatly increase the bandwidth of a dual-emission filter set since the emission spectra can overlap. In this case, increasing the width of the excitation wavelength is the better option.

4.1.3 Excitation intensity

The excitation intensity should be sufficient to give adequate signal, but not to bleach the indicator for the period of the experiment. The excitation should only be increased as a last resort. Remove any neutral density filters (these filters can be thought of as grey filters, with equally reduced transmission across all wavelengths) in the excitation light path and ensure the arc lamp is centred properly (consult your microscope manual). Increase the bandwidth of the excitation filter if possible.

4.2 Image too bright

If the signal from the specimen is too bright for the camera, the detector will become saturated, i.e. the signal should be, say, around 300, but the (8-bit) camera image can only deal with 256 values. Often, saturation occurs at the peak of a Ca^{2+} response, by which time the experiment is part-way through. A few preliminary experiments where naïve cells loaded with an indicator are exposed to a Ca^{2+} ionophore and high extracellular Ca^{2+} will provide a good indication of the expected maximum signal.

4.3 Indicator bleaching

Photobleaching of the indicator is one of the biggest problems in fluorescence microscopy, and is a common problem with observations of Ca^{2+} signals over prolonged periods. Bleaching can affect Ca^{2+} changes reported by both single- and dual-wavelength indicators. Bleaching occurs when an excited fluorophore, instead of emitting a photon and returning to a non-excited state, reacts with molecular oxygen to form a non-fluorescent molecule. Scavengers of reactive oxygen species (ascorbate, Trolox) have been used to reduce bleaching (4), although such scavengers may have biological effects. Bleaching of the indicator follows an exponential decay and may be corrected for if a sufficient control period is allowed for calculation of the decay constant.

5 Single-photon laser scanning confocal microscopy

5.1 General introduction

Photometry and video imaging only offer limited information about the spatial distribution of Ca^{2+} inside single cells. The reason for this lies in the fact that PMTs and video cameras collect the fluorescence originating from a thick (> 5 μm) focal plane within the specimen. In contrast, confocal microscopes can 'optically slice' living or fixed cells by only capturing emitted light from thin optical sections (~ 1 μm or less). This effect is achieved by introducing a spatial filter in form of a pin-hole or slit aperture that blocks emitted light from out-of-focus sections. For a detailed description of the principles of confocal microscopes we refer the reader to the excellent text edited by Pawley (5).

5.2 General arrangement of a laser scanning confocal microscope

The general set-up of a confocal microscope consists of four major components: (i) a laser, (ii) a scan head, (iii) a microscope, and (iv) a computer to control hardware function and capture images. Lasers are almost exclusively used for confocal microscopy due to the need for intense illumination. Different types of lasers are employed, depending on the particular excitation wavelength required. Specially tuned argon ion lasers are optimized for UV emission, with lines usually available at 351 nm and 364 nm and output energies of a few hundred milliwatts. For visible indicators, lower power argon ion lasers, with output energies of several tens or more of milliwatts, offer excitation wavelengths at 457 nm, 488 nm, and 514 nm. Longer wavelengths can be obtained from argon-krypton (488 nm, 568 nm, 647 nm) or He-Ne (534 nm, 633 nm) lasers.

The scan head is responsible for transmitting the excitation light to the specimen on the stage of the microscope. Usually the laser beam is fed directly into the scanning head by means of a light guide, and after passing through an excitation filter the laser beam is scanned horizontally and vertically across the field of view. Scanning is achieved either with two galvanometer-driven mirrors (one for horizontal and one for vertical scanning) or with a single mirror and a device known as an 'acusto-optical deflector' (AOD) (the mirror performs the slow vertical scanning while the AOD is responsible for the fast horizontal scanning). The emitted light is collected simultaneously with the excitation of the specimen, and is usually detected by a PMT. In video imaging, a complete image of the excited specimen is generated on the CCD chip. However, with laser scanning confocal microscopy, the intensity of the fluorescence signal at each point in the field of view is used to construct an image by the computer (but see the Nipkov disc technology described below).

5.3 Spinning disc confocal microscopes

One of the few commercially available confocal microscopes that is able to generate a direct confocal image uses a Nipkov disc. The specimen is illuminated by light that passes through thousands of pin-holes arranged in a geometrically precise pattern (e.g. an Archimedes spiral) on a rapidly spinning disc. The effect is that all parts of the specimen are illuminated quasi-simultaneously. The emitted light passes back through the same pin-holes, thus excluding out-of-focus light originating from other points of the specimen. The emitted light is not de-scanned as with other confocals, but instead forms a real time colour image that can be viewed via an eyepiece or camera. This type of system can also be practically used with non-laser light sources, thus expanding the range of usable fluorophores. Disc spinning devices are considerably faster than real laser scanning confocal microscopes but usually exert lower detection efficiencies. A recent development has tried to address the latter problem by placing micro-lenses in front of each hole in the disc and thereby collecting more light (Yokogawa, Japan). Another problem inherently associated with spinning discs is the lack of available high-sensitivity CCD cameras with a sufficient quantum efficiency in the green part of the spectrum to allow for fast data acquisition with a good optical resolution. Due to the advantages of on-chip integration (see Section 3.2) spinning disc confocal microscopes are particularly useful for fluorescence detection that is not time critical, such as green fluorescent protein work and fixed specimens.

6 Advantages and pitfalls of confocal microscopy

Confocal microscopy offers high spatial and temporal resolution fluorescence measurements in living cells. The point-spread function (PSF) describes how an ‘infinitively’ small light source (point source) is represented by the optics and defines the optical performance of the microscope. Typical PSF dimensions are 300 nm•300 nm•600 nm (X•Y•Z direction) for an objective with a high numerical aperture (NA $>$ 1.2), depending on the type of spatial filter used (i.e. pin-hole versus slit aperture). Microscopes with a pin-hole achieve the best optical resolution. However, the temporal resolution critically depends on the scanning mechanism used.

6.1 Speed considerations

With current galvanometer-driven mirror technology, the time taken for scanning a full frame ($\geq 512 \times 512$ pixels) is around one second or longer. This speed

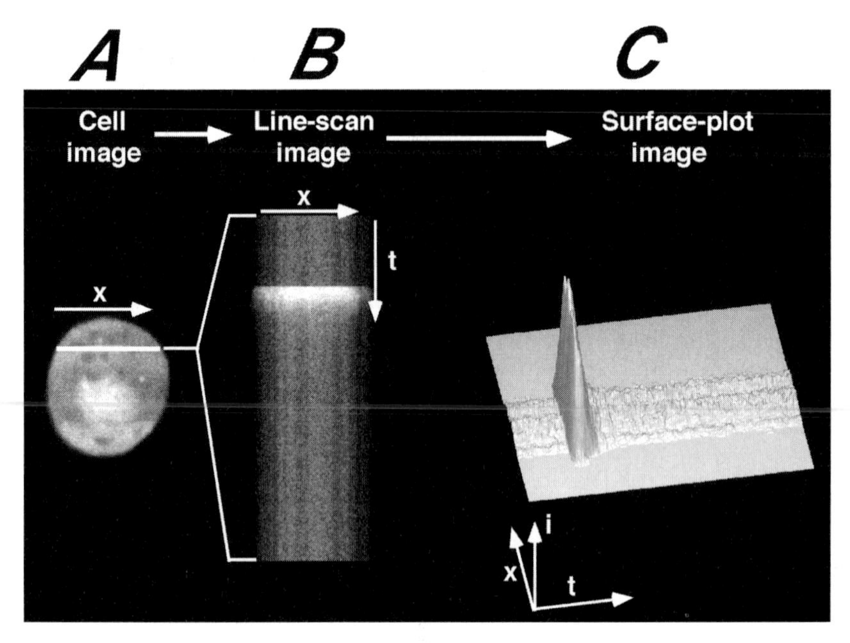

Figure 1 Different modes of acquisition and representation of confocal data. (A) An x/y-image of a fluo-3 loaded cultured ventricular myocyte from a neonatal rat at rest. Restriction of the scanning process to a single line (marked in white) results in the generation of a line scan image as depicted in (B), whereby the preserved spatial dimension runs horizontally while the time axis runs vertically. This cell was field-stimulated and the evoked Ca^{2+} transient can be seen as a horizontal band of higher brightness. Coding of the cellular fluorescence in the height of a 3D surface results in a so-called surface plot (C). The figure was modified from ref. 10.

is suitable for following slow Ca^{2+} oscillations or slow Ca^{2+} waves, but will not allow faster events to be resolved. With these types of microscopes the time-resolution can only be enhanced by either decreasing the frame size or by performing line scan imaging. The former approach decreases the optical resolution, although typical frame sizes with an acceptable optical resolution ($\sim$ 100 $\times$ 100 pixels) can be acquired with rates of up to $\sim$ 8 Hz. For line scan imaging, one spatial dimension is sacrificed to improve on the temporal resolution. Essentially, scanning is performed in one direction only (see *Figure 1*). Usually, the scanned line has to run horizontally across the field of view, although some confocal microscopes have the ability to rotate the angle of the scan line. By using motorized mirrors for line scanning, the maximal rate of line acquisition approaches 1000 lines/sec.

AOD-driven confocal microscopes or Nipkov disc-based systems can achieve much higher frame rates (up to several hundreds frames per second). At these rates the integration time per pixel is very short ($\sim$ 100 nsec) and consequently the noise can be significant. Typical acquisition rates for these types of microscopes are somewhere between 10 and 120 frames per second. The speed can even be increased further for AOD-driven confocals by only using the AOD for horizontal line scanning. In this mode, AOD-driven microscopes can achieve several ten of thousands of lines per second. Due to their operating principles, confocal microscopes using Nipkov discs do not allow a line scanning mode.

6.2 Problems with changing cell shape

A significant problem when monitoring Ca^{2+} signals in living cells with confocal microscopy occurs when the cells change their shape. In this situation, changes of fluorescence in the plane of focus cannot unequivocally be ascribed to changes of Ca^{2+} in a particular volume element (voxel) in the cell. This is because the change of shape (e.g. due to migration or contraction) causes voxels to move into and out of the plane of focus. This problem is particularly prominent when working with single-excitation single-emission indicators. The situation can partially be accounted for by either using ratiometric indicators or combining two single-excitation single-emission indicators (see Section 2.2.2).

6.3 Intensity depth, gain, and off-set

Many of the problems encountered during video imaging and photometry, e.g. setting the correct black level, are also faced in confocal imaging. Most commercially available confocal microscopes are 8-bit systems. As discussed above (see Section 3.7) 256 grey levels can be rather restrictive, and it is necessary to adjust the gain (contrast) and off-set (brightness or black-level) settings to allow most of the grey levels to be used.

Fluo-3, for example, has a dynamic range of ~ 10 inside cells (i.e. a tenfold increase in fluorescence output upon binding Ca^{2+}). Consequently, it is important not to begin an experiment with gain setting that will cause the PMT to become saturated at the peak of a Ca^{2+} response. On the other hand it is not very useful to start with gain settings that keep fluorescence values close to zero, since the percentage error inherently associated with small changes in fluorescence and arising from digital noise is considerable. The brightness setting also has an important bearing on the range of grey levels that can be utilized by the confocal microscope.

Let us consider this with a simple example to illustrate the effect of gain and brightness settings. In a hypothetical specimen the 'real' fluorescence undergoes a doubling of its value, say from a resting value of 30 to a peak response of 60. Changing the gain will alter the absolute values so that the resting fluorescence will increase, for example to 50 and the peak response to 100, but the relative change (twofold) is still preserved.

The situation is different when the brightness setting is changed. Considering our hypothetical example, where the resting and peak intensities were 30 and 60 respectively. If we decrease the brightness by 15, then the resting intensity becomes 15, whilst the peak response is now threefold higher at 45. On the other hand a positive shift of the brightness will add to both fluorescence values and subsequently decrease the relative fluorescence change.

In summary, the gain or contrast setting amplifies the fluorescence values

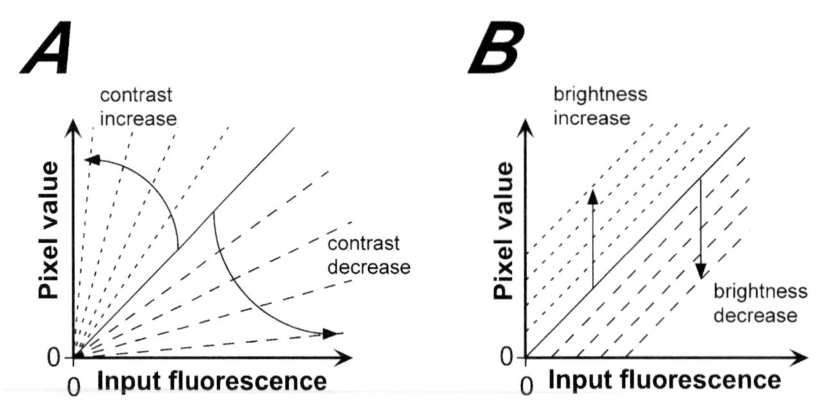

Figure 2 Effect of varying the contrast (or gain) and brightness (or off-set) on the resulting fluorescence. (A) Changes to the resulting fluorescence image (pixel value) when the contrast is either increased or decreased. (B) Equivalent changes when the brightness is varied.

but preserves relative changes of the 'real' fluorescence, while the brightness setting shift the recorded values vertically and falsifies relative fluorescence changes. The effect of changes to gain and brightness settings on pixel intensity values are shown in *Figure 2*. The situation becomes even more complicated in that gain and brightness settings affect each other (see also ref. 2). The ideal situation for imaging of any kind is to start with a completely neutral off-set, i.e. a brightness position that results in a pixel value of zero in background areas of the field of view. In practice this is difficult to achieve. It is therefore much more advisable to start with a clear positive off-set of ~ 10 grey levels which is subtracted from the data off-line during the analysis.

Despite the large variety of confocal microscopes, there is a sequence of routine steps to be followed prior to starting an experiment, which are common to all confocal devices. The standard protocol used in the authors' laboratory is described in *Protocol 3*.

Protocol 3

Preparation for confocal microscopy

Equipment and reagents

- Confocal microscope
- See Protocol 2

Method

1 Switch on the laser, control tower, scanning head, and the computer controlling the confocal microscope. This might sound very trivial, but again and again one of these components does not get switched on and then the starting-up procedure will be interrupted or no fluorescence image will be obtained. Modern argon ion lasers

Protocol 3 continued

visible confocal microscopy only need very few minutes for warming up and stabilizing. Argon ion lasers used for UV confocal microscopy should be warmed-up for approx. 5–10 min.

2 We assume that the cells have been loaded as described in Protocol 2 and the coverslip is mounted in an appropriate chamber.

3 Mount the chamber on the stage of the confocal microscope and bring the cells approximately into focus in the transmission mode of the microscope.

4 Start-up the control software.

5 Start the scanning process and after visualizing the cells reduce the laser settings to a minimal value so that you can just recognize the cells.

6 Adjust the contrast (also called 'gain' on some systems) to a value that you have determined previously to allow imaging of the cells (i.e. using most of the 8-bit space).

7 Choose (or generate) a LUT (look-up table) that clearly marks pixels with very low fluorescence values (< 5) in a colour that is readily available.

8 Adjust the brightness (sometimes called 'black level') to the lowest possible value so that the 'cell-free' areas of your image just do not show any pixels with undetectable fluorescence.

9 Finally adjust the laser intensity to a level which allows easy identification of the individual cells.

7 Examples

The following section presents some example protocols for confocal imaging of Ca^{2+} signals, and also some possibilities for the analysis and presentation of these confocal images. In addition, problems in the interpretation of confocal data are discussed.

7.1 Linear Ca^{2+} waves in image mode

This is one of the easiest types of Ca^{2+} signal to study using confocal microscopy. After loading the cell and setting up the confocal microscope, the most important decision is to choose and appropriate scanning speed. Since the velocity of Ca^{2+} wave propagation is dependent upon the cell type, there are no universal settings. However, image capture rates of at least 10 Hz are preferable.

An example of such a linear Ca^{2+} wave is shown in *Figure 3Aa*. This sequence was obtained at an acquisition rate of 8 frames per second on an Bio-Rad MRC-600. 12 consecutive ratiometric images of a fluo-3/fura red loaded guinea-pig ventricular myocyte are shown. The propagation of Ca^{2+} waves from the bottom to the top of the cell can be seen as bands of white intensity in the grey scale image. The same myocyte visualized a couple of minutes later showed a different pattern of response (*Figure 3Ba*). Instead of Ca^{2+} wave propagating linearly

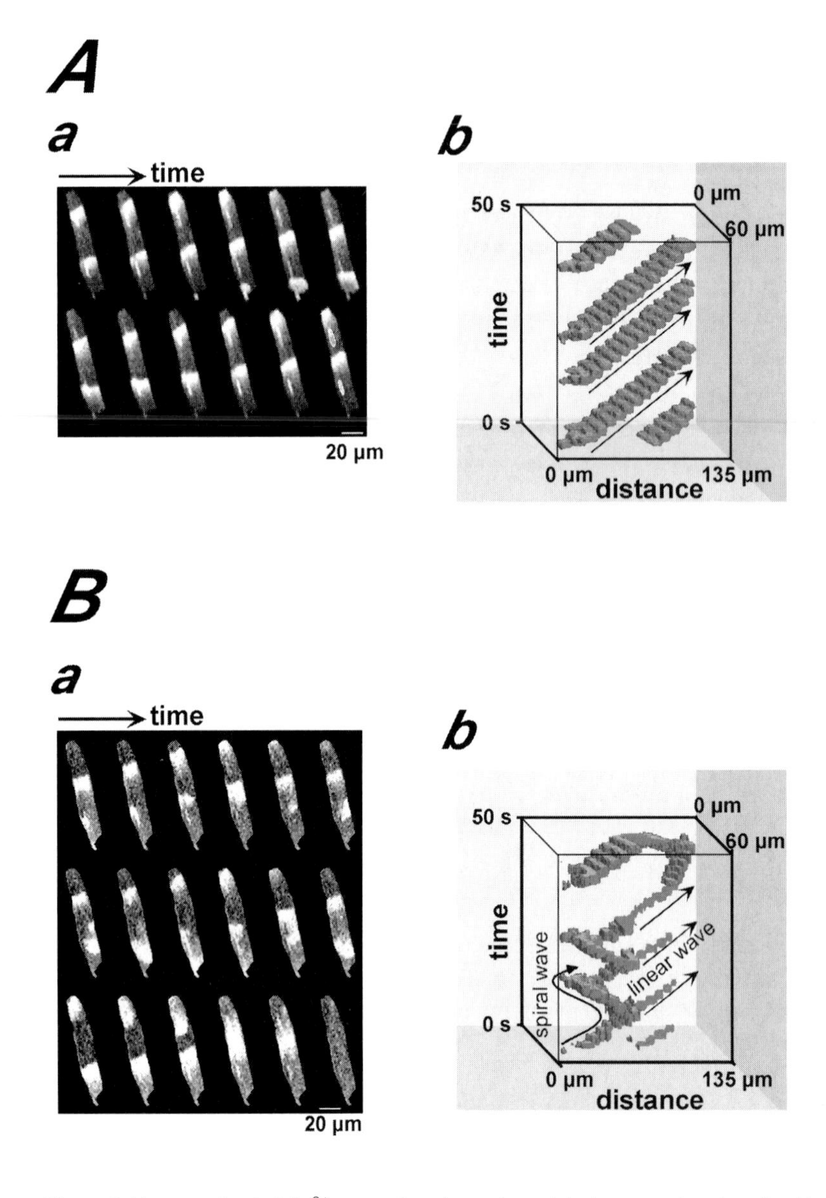

Figure 3 Linear and spiral Ca^{2+} waves in guinea-pig ventricular myocytes visualized by ratiometric confocal microscopy. Guinea-pig ventricular myocytes were loaded with fluo-3 and fura red simultaneously. Sequences of ratiometric confocal images at a rate of ~ 8 frames/sec are shown in (Aa) and (Ba) whereby time runs from upper left to lower right. (A) The time course of spontaneously occurring linear Ca^{2+} waves. Stacking such images on top of one another and rendering volume elements which display a resting Ca^{2+} level transparent results in a pseudo 3D object (Ab and Bb; time runs from the bottom to the top). (Ab) The 3D representation of linear Ca^{2+} waves (as elongated 'cigar'-like objects). (Bb) An additional spiral Ca^{2+} wave (as a 'corkscrew'-like object in the lower left corner). Propagation directions of the waves are marked with arrows. The panels were modified from ref. 6.

from bottom to top, spiral Ca^{2+} waves are observed revolving around the persistently bright nucleus.

Although the linear and spiral Ca^{2+} waves can be seen in the montage of grey scale images in *Figures 3Aa* and *3Ba*, the different patterns of response become strikingly clearer when they are transformed into a 3D representation (*Figures 3Ab* and *3Bb*). From the plot in *Figure 3Bb* the spiral Ca^{2+} wave is apparent (corkscrew arrow) as well as several linear waves descending from the spiral wave (straight arrows). This example illustrates the advantage of utilizing different forms of data representation. For additional directions on how to construct such pseudo-3D images, we refer the reader to the original paper (6).

7.2 Linear Ca^{2+} wave in line scan mode

In situations where the image mode is too slow to resolve rapid Ca^{2+} changes the next best option is to switch to line scanning. *Protocol 4* describes the basic steps involved in capturing line scan images. An example where line scanning revealed spatially-distinct Ca^{2+} changes that would have been missed by the slower image mode are shown in *Figure 4*. In this example a Ca^{2+} wave can be seen to propagate through the cytoplasm of a fluo-3 loaded HeLa cell. The grey scale representation shows that the Ca^{2+} wave propagation was not smooth but consisted of saltatoric or 'step-like' increments (7).

Although line scanning provides the most rapid means of studying Ca^{2+} changes, some care has to be taken in the interpretation of such images. In particular, the calculation of wave velocity can be overestimated depending upon the position of the scan line relative to the direction of wave propagation. Intuitively, one would calculate the wave speed as the distance along the line

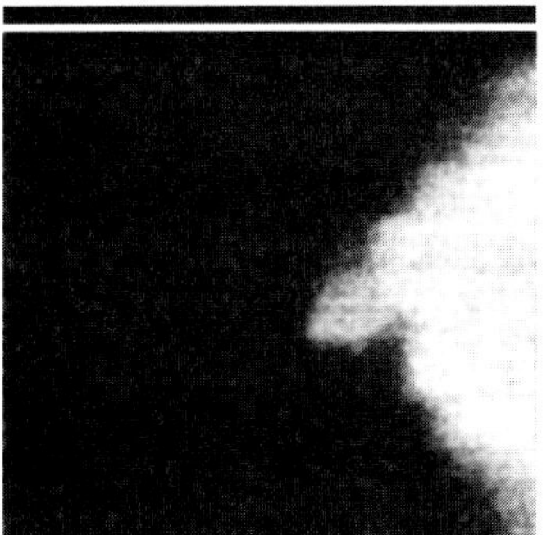

Figure 4 Line scan image of a Ca^{2+} wave in a histamine-stimulated HeLa cell. The HeLa cell was loaded with fluo-3 and line scans were acquired at a rate of ~ 100 lines/sec along the longitudinal axis. Stimulation of the cell with histamine evoked a propagating Ca^{2+} wave (as indicated by the 'V'-shape wave front). This Ca^{2+} wave did not display a smooth wave front instead it can be seen that the waves 'jumped' between locations of Ca^{2+} release while propagation between these locations was slow. The image was modified from ref. 7.

travelled by the wave in a given time. However, this is only true if the wave front travels perpendicular to the direction of the scan line. If the scan line is not at 90 degrees to the direction of propagation, then the apparent velocity will increase, as illustrated in *Figure 5*.

Protocol 4

Setting up for line scanning

1. Get a confocal image of your cells on the screen of the computer.
2. Consult the software manual for your confocal microscope and start the line scan software.
3. Choose a suitable line along the specimen for imaging.
4. Adjust the speed of recording, bearing in mind that rapid line scanning may cause acute bleaching of the indicator and worsen the signal-to-noise ratio.

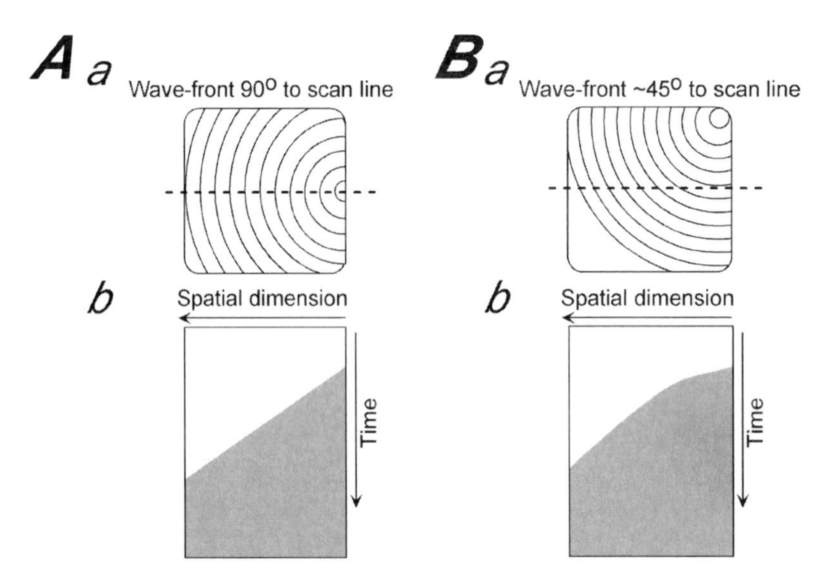

Figure 5 The apparent characteristics of a propagating Ca^{2+} wave depend on the line scan/wave front angle. (A) The propagation of a Ca^{2+} wave in a single cell (Aa) when the line chosen for the line scan image (dashed line in Aa) has a 90° angle to the wave front. The resulting line scan image (Ab) shows the expected smooth wave front. (Ba and Bb) When the chosen line deviates from this 90°, the resulting line scan image shows a clearly distorted wave front.

7.3 Elementary Ca^{2+} signals

7.3.1 Background

Ca^{2+} waves and oscillations are considered to be 'global' signals, since they usually invade the entire cell, including cytoplasmic and nucleoplasmic compartments, and some organelles, e.g. mitochondria. In recent years, there has been a growing awareness that Ca^{2+} is often employed as a local signal in a wide variety of cell types (1).

Cells utilize two sources of Ca^{2+} for generating signals; Ca^{2+} release from intracellular stores and Ca^{2+} entry across the plasma membrane. Both Ca^{2+} release and entry mechanisms can give rise to local or 'elementary' Ca^{2+} signals. Examples of elementary Ca^{2+} signals are the Ca^{2+} puffs in *Xenopus* oocytes and HeLa cells (7, 8), and the Ca^{2+} sparks in cardiac myocytes (9–11). Since much interest is presently focused on these elementary Ca^{2+} signals, the following section illustrates how they can be recorded using confocal microscopy.

7.3.2 Line scan or image mode?

Elementary Ca^{2+} signals are brief, for example Ca^{2+} sparks have a time to peak of ~ 10 msec and decay within 100 msec. The rapid up-stroke and recovery of elementary Ca^{2+} signals necessitates the use of a rapid scanning, otherwise the such events will not be resolvable. For confocals based on galvanometer-driven mirrors the only solution is to use line scanning. An example of Ca^{2+} sparks detected using line scanning of a fluo-3 loaded ventricular myocyte is depicted in *Figure 6*. In order to obtain a high temporal resolution consecutive lines were acquired at 500 lines per second (i.e. 2 msec per line) using a Bio-Rad MRC 600 system. The grey scale image in *Figure 6* shows that when the myocyte was depolarized there were distinct inhomogeneities in the resulting Ca^{2+} signal.

Using line scanning to study Ca^{2+} sparks in a cell type such as a cardiac myocyte is helped by the fact that myocytes have an oblong shape and the position of the scan line is not crucial—it will almost always transect a Ca^{2+} spark site. In cells that have a less regular shape, line scanning may be too limited. In the example of a differentiated PC12 cell shown in *Figure 7*, the neuritic tree is branched and highly irregular (*Figure 7B*). Choosing only one of the neuritic branches for line scanning would discard much information. Instead, this sample was scanned at 30 Hz in image mode using the AOD-based Noran Oz confocal microscope. Using the image mode allowed the identification of discrete subcellular regions that responded to either caffeine (*Figure 7A*) or bradykinin (*Figure 7C*) (12).

7.3.2 Ca^{2+} indicator

Since elementary Ca^{2+} signals have only a fraction of the amplitude of a global Ca^{2+} wave, it is helpful to use an indicator (an appropriate gain settings) that allows such events to be discriminated from the background. In our hands, fluo-3 and fluo-4 are the most suitable indicators. Their nearly tenfold change in emission upon binding Ca^{2+} means that the background signal is low and regions with elevated Ca^{2+} can be easily detected. Furthermore, we find that fluo-3 has a sufficient dynamic range to enable it to detect both elementary Ca^{2+} signals and global Ca^{2+} waves using the same laser intensity, brightness, and gain settings.

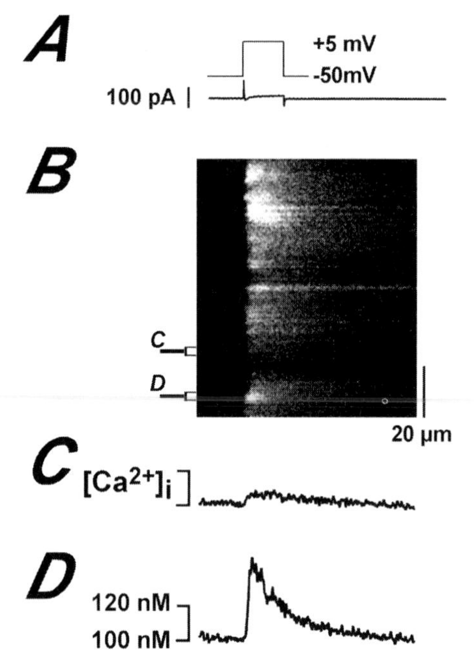

Figure 6 Ca^{2+} current-triggered Ca^{2+} sparks in guinea-pig ventricular myocyte. A single guinea-pig ventricular myocyte was patch clamped and loaded with fluo-3. L-type Ca^{2+} currents were inhibited by almost 95% by the application of 10 μM nifedipine. Step depolarizations of the membrane potential from –50 mV to +5 mV elicited a small residual L-type Ca^{2+} current (A) that was still able to evoke a substantial Ca^{2+} release signal (as seen in the line scan image in B). Comparison of two representative areas along the scanned line showed that the Ca^{2+} transient consisted of localized Ca^{2+} sparks (C) intermitted with regions of only small Ca^{2+} changes (D). The figure was modified from ref. 14.

8 Multi-photon confocal microscopy

8.1 Background

In comparison to single-photon microscopy (see above) that achieves the confocality by introducing a pin-hole or slit as a spatial filter on the emission side, confocal microscopy using multi-photon molecular excitation (MPME) generates a 'confocal' excitation, i.e. an excitation that is confined to a very thin layer in the specimen. This effect is based on the necessity to absorb two or more photons quasi simultaneously. The required photon density can only be found in a very thin layer in the centre of the focal plane. The basic mechanism behind MPME is that the energy for excitation is not delivered in a single package (or photon) of high energy. Instead, the excitation energy is supplied through the simultaneous absorption of for example two photons, each delivering half of the necessary excitation energy. Even from this basic description, some of the advantages of MPME over single-photon microscopy are apparent. First, for excitation of dyes which usually require UV light, and therefore introduce optical problems for confocal microscopy, such as chromatical aberration, excitation

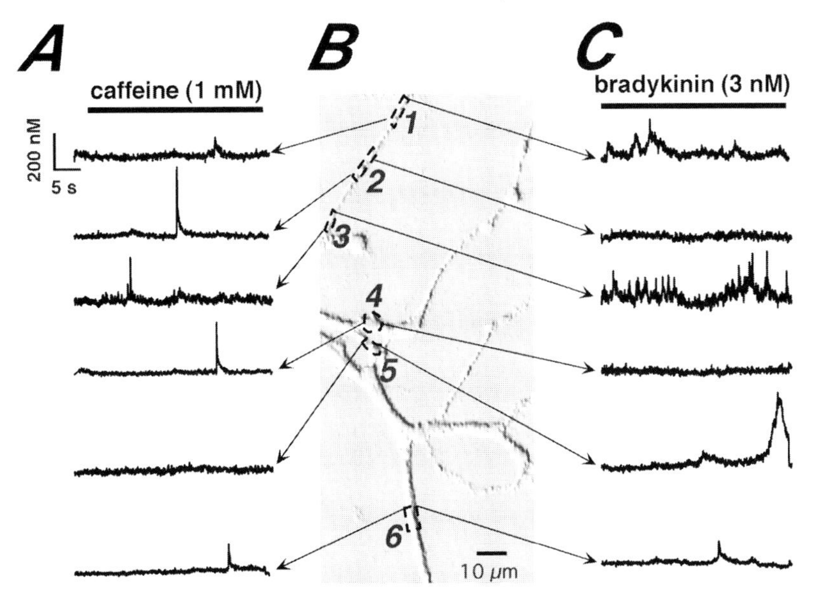

Figure 7 Caffeine- and bradykinin-evoked elementary Ca^{2+} release signals in a NGF-differentiated PC-12 cell. (A and C) Localized Ca^{2+} signals obtained after caffeine (A) or bradykinin (C) stimulation of a fluo-3 loaded NGF-differentiated PC-12 cell. (B) The neuritic tree and the locations from which Ca^{2+} signals were averaged. Whereas some regions displayed responsiveness only to one of the agonists (regions 2 and 4) the other locations responded to caffeine (recruiting ryanodine receptors) and bradykinin (recruiting inositol 1,4,5-trisphosphate receptors). This figure was modified from ref. 12.

light of double the wavelength can be used. For example, the 350 nm light needed for Indo-1 excitation can be obtained with MPME using a wavelength of 700 nm (in the case of two-photon excitation). Another advantage in using longer wavelengths is that most microscope objectives are optically corrected for the use of different visible wavelengths; an adequate correction for the use of UV light is usually lacking. For a more detailed description of the problems associated with the use of UV light in confocal microscope can be found in Niggli *et al.* (13). Secondly, since MPME results in the generation of 'confocal' excitation all the emitted light originates from a confocal plane and a spatial filter (either pin-hole or slit aperture) is not necessary. Thirdly, red light is much less absorbed and scattered by living tissue than blue. Thus the maximal depth of penetration for MPME light is significantly increased. This is a particular advantage when thick tissue sections are used. With one-photon confocal microscopy the depth of penetration is usually limited to a few tens of μm. MPME will extend this range by at least an order of magnitude.

8.2 Hardware

In principle, MPME confocal imaging is performed under similar conditions and constraints to single-photon imaging. The same care must be taken to set brightness and gain levels as described above for the standard single-photon confocal microscope. A major difference between single-photon confocal microscopy and MPME confocal imaging is the type of laser used. The most versatile sources for this purpose are Ti:Sapphire lasers. These lasers can be tuned from around 680 nm to well above 1100 nm. In some cases it may be possible to switch from single-photon imaging to MPME by simply replacing the laser.

9 Multi-photon photolysis

Although MPME has advantages over single-photon confocal microscopy, the cost and complexity of multi-photon systems means that they may not always be most appropriate. Another powerful application of MPME is photorelease of caged compounds in diffraction-limited volumes. Caged compounds represent inert photo-labile forms of ions, buffers, and biologically active molecules. Exposure of these compounds to intense UV light immediately liberates the ion or molecule from the chemical cage.

Caged compounds can be uncaged using UV sources such as flash lamps and lasers. However, although these sources can be focused within a cell, they produce a 'cone' of excitation where photolysis occurs above and below the desired focal plane. With MPME photolysis only occurs in the cellular volume where the photons of light arrive with sufficient frequency. In practice, MPME can be used to uncaged compounds in volumes smaller than one μm^3. Due to the fact that only a tiny volume in the cytosol is excited these photolysis events can be reliably reproduced many times without substantial consumption of the caged precursor. An example for such an MPME-generated local Ca^{2+} response is shown in *Figure 8*.

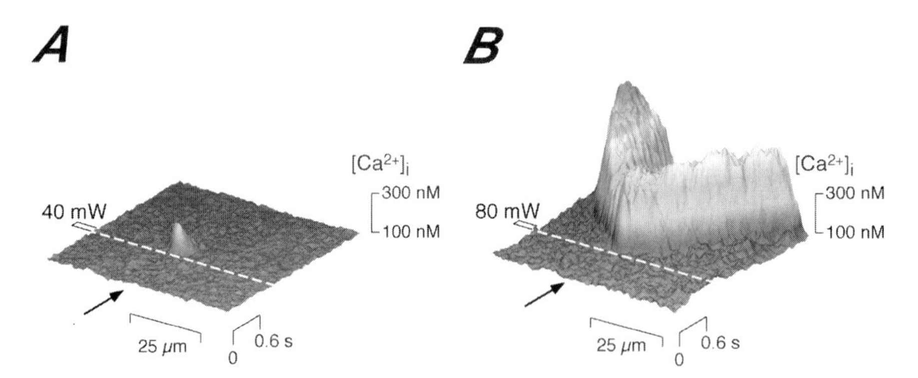

Figure 8 Diffraction-limited photolysis of caged Ca^{2+} by MPME in a guinea-pig ventricular myocyte. The cardiac cell was patch clamped and loaded with fluo-3 and caged Ca^{2+} (DM-nitrophen). The beam of a Ti:Sapphire laser (MIRA, COHERENT) was focused into the myocyte. The laser beam had a wavelength of 710 nm and an energy of 40 mW (A) or 80 mW (B). While the low-energy beam elicited only a localized Ca^{2+} signal (A), the higher-power photolysis event triggered a propagating Ca^{2+} wave (B). In both (A) and (B), the onset of photolysis is indicated by the dashed lines, while the brackets at the left side of the line scan images depict the excitation duration. The figure was modified from ref. 15.

10 Data processing

10.1 Calibration of the indicator fluorescence

The equations given below can be used to convert fluorescence intensities (or ratios) into Ca^{2+} concentrations. The fluorescence response of an indicator to different Ca^{2+} concentrations is determined by its maximum and minimum fluorescence (F_{max} and F_{min} respectively), and the indicators dissociation constant (K_d) with Ca^{2+} by the equations:

$$[Ca^{2+}] = K_d\left(\frac{F - F_{min}}{F_{max} - F}\right)$$

Single-wavelength equation

Note: This equation is suitable for the single-wavelength indicators that increase their fluorescence upon calcium binding (the vast majority of indicators belong to this type). For such indicators F_{min} corresponds to the calcium-free form of indicator and F_{max} corresponds to the calcium-bound form.

$$[Ca^{2+}] = K_d\left(\frac{R - R_{min}}{R_{max} - R}\right)\left(\frac{F_{f,\lambda2}}{F_{b,\lambda2}}\right)$$

Dual-wavelength equation

The values for F_{min} and F_{max} (or R_{min} and R_{max}) are measured by adding a calcium ionophore (e.g. 10 μM A23187 or ionomycin) in the presence of excess EGTA (e.g. 10 mM) or excess $CaCl_2$ (e.g. 10 mM). The ratio values (R) are, by convention, calculated by dividing the fluorescence intensity at a wavelength corresponding to the Ca^{2+}-bound indicator (λ1, e.g. 340 nm for fura-2) with the fluorescence at a wavelength corresponding to Ca^{2+}-free indicator (λ2, e.g. 380 nm for fura-2). Ratio values are used in the dual-wavelength equation rather than the fluorescence values. Similarly, the ratio at maximum and minimum Ca^{2+} concentration (R_{max} and R_{min}) are used instead of fluorescence max and minima. The ($F_{f,\lambda2}/F_{b,\lambda2}$), factor is calculated using measurements of fluorescence of calcium-free form of indicator ($F_{f,\lambda2}$) and fluorescence of calcium-bound form of indicator ($F_{b,\lambda2}$). For a more detailed explanation of how to calibrate fluorescence measurements see *Protocol 5*.

Protocol 5

Simple calibration of Ca^{2+} concentration in fura-2 loaded cells post-experiment

This calibration is suitable for experiments in which fluorescence is measured photometrically and cells can be perfused with extracellular solution.

1. It is assumed that an experiment has taken place on fura-2 loaded cells, and that the ratio of fluorescence at 340 nm excitation and 380 nm excitation has been monitored throughout.

Protocol 5 continued

2 At the end of the experiment, incubate the cells in a solution containing calcium ionophore (e.g. ionomycin, A23187; 1–40μM depending on the cell type; 10 μM is sufficient for most cell types). This solution should be calcium free and should also contain a calcium chelator (e.g. 1–10 mM EGTA). Note that A23187 is fluorescent and its contribution to the background signal should be taken into account. Alternatively, the non-fluorescent 4-bromoA23187 can be used. Preliminary experiments should be conducted to estimate the time needed to deplete cellular calcium, 20–40 min is usually sufficient for most cell types. It is useful to slightly alkalinize the extracellular solution (e.g. to pH 7.6–7.8). This increases the efficiency of the ionophore, which is particularly important for transport of calcium at low calcium concentrations.

3 Select appropriate wavelengths for excitation (e.g. 340 nm and 380 nm) and emission (e.g. 510 nm). Start measurements in the EGTA and ionomycin containing solution. Determine parameters: $F_{f,\,380}$ (fluorescence of the calcium-free indicator at 380 nm excitation), $F_{f,\,340}$ (fluorescence of the calcium-free indicator at 340 nm excitation).

4 Calculate R_{min} as:

$$R_{min} = F_{f,\,340} \,/\, F_{f,\,380}$$

5 Change extracellular solution to one containing Ca^{2+} at a concentration of 5–10 mM. This solution should contain the same concentration of the ionophore and be at the same pH as the calcium-free solution.

6 Observe a sharp decrease of fluorescence at 380 nm excitation and increase of fluorescence at 340 nm. After application of high calcium solutions, many cell types will start to swell—it is advisable at the beginning of the experiment (step 3 of this protocol) to select the window for measurements of fluorescence slightly larger (at least an extra 20% in linear dimensions) than the image of the cell.

7 Wait until the intensity of both signals has stabilized (typically 5–15 min). At that time measure $F_{b,\,380}$ (fluorescence of the calcium-bound form of the indicator at 380 nm excitation), and $F_{b,\,340}$ (fluorescence of the calcium-bound form of the indicator at 340 nm excitation).

8 Use the measured parameters to calculate R_{max}:

$$R_{max} = F_{b,\,340} \,/\, F_{b,\,380}$$

9 Calcium concentrations can be calculated using the dual-wavelength equation (see above) for fura-2 derived by Tsien and colleagues (16):

$$[Ca^{2+}] = K_d \left(\frac{R - R_{min}}{R_{max} - R}\right)\left(\frac{F_{f,\,380}}{F_{b,\,380}}\right)$$

10 Typical dissociation constants (K_d) for fura-2 are 135 and 224 nM at 22 °C and 37 °C respectively (see Molecular Probes Catalog (http://www.probes.com) for values of other indicators). $R = F_{340}/F_{380}$; the ratio of fluorescence using 340 and 380 nm excitation determined at different time points during the experiment.

The calibration procedure outlined above assumes that the contribution of autofluorescence is negligible. Separate test experiments can be done to verify this point:

(a) Compare the intensity of fluorescence of unloaded cells with fluorescence of cells loaded with indicator.

(b) Use Mn^{2+} to quench the fluorescence of fura-2 to reveal the contribution of autofluorescence. This can be done by changing the extracellular solution after step 7 in *Protocol 5* to a solution containing ionophore and 2 mM of $MnCl_2$ (this solution should be nominally calcium free with no EGTA added).

Particular attention should be given to the comparison of $F_{b,\,380}$ fluorescence of the cell loaded with fura-2 when the cytoplasm is saturated with calcium and $F_{380a.f.}$ (autofluorescence of the cells at 380 excitation determined either in unloaded cells or after Mn^{2+} quench of the indicator fluorescence). The contribution of autofluorescence at this wavelength of fura-2 excitation can be considerable even in cells containing a 100 μM fura-2. The error due to autofluorescence is most significant (for fura-2 measurements) at the peak of calcium responses, especially when the calcium concentration in cells increases beyond 1 μM. The correction can be obtained by subtracting the values of autofluorescence from the fluorescence measurements obtained at both wavelengths of excitation during the experiment. The calibration of single-wavelength indicators can be performed in a similar way to two-wavelength indicators.

Another potential artefact of calcium measurements (occurring due to binding of transitional metals and zinc to calcium indicators) is considered by V. Snitsarev and A. Kay (Chapter 3) in this volume.

Estimates for the affinities of indicators for Ca^{2+} are usually provided by the manufacturer. However, K_d values vary with temperature, pH, viscosity of the medium, protein binding, and the presence of other ions. It has been shown that the K_d of an indicator can vary between cell types, and even subcellular locations. Therefore, for accurate measurements of Ca^{2+}, it is often preferable to empirically measure the K_d by permeabilizing cells using a calcium ionophore (e.g. 10 μM A23187 or ionomycin) and monitoring changes in fluorescence following application of external solutions containing known Ca^{2+} concentrations.

10.2 Self-ratio of single-wavelength indicators

In order to circumvent some of the problems associated with single-wavelength indicators, a 'self-ratio' may be calculated. This procedure assumes that at rest there are no Ca^{2+} gradients within the cell and that any variations in signal intensity are due to variations in indicator distribution. By dividing all images by the average of the 10–20 images at rest, uneven indicator loading is corrected for in the final data. This does not correct for changes in indicator distribution during the experiment, i.e. compartmentalization, leakage, or bleaching.

10.3 Pseudocolour

As mentioned previously, the human eye can discriminate less than 50 grey levels, so to enhance small changes in grey levels, pseudocolours may be used. However, pseudocolour can be misleading. Dependent upon the LUT used (see Section 3.5), small changes in intensities may be masked by assigning large ranges of grey levels to be the same pseudocolour. Furthermore, dramatic changes in the pseudocolour may actually only represent a small change in actual intensity. Small differences in grey scales may be made more evident, whilst not deceiving the observer, by simply increasing the contrast of the image. This also helps to reduce publication costs!

10.4 Archiving

An important consideration with regard to data is that of long-term storage. With photometry this is not much of an issue since even the longest experiment generates a relatively small data file. However, imaging (video or confocal) generates vast amounts of data. Taking an average example, a 30 minute experiment capturing one 768 × 512 8-bit (i.e. 8 bits = 1 byte) image every 2 sec generates a file 354 Mb in size. Of the currently available options for digital data storage today, the best is CD ROM. This is the cheapest form of data storage in terms of cost per megabyte of data, ease of data access and cross-platform transferability.

References

1. Berridge, M., Bootman, M., and Lipp, P. (1998). *Nature*, **395**, 645.
2. Lipp, P. and Niggli, E. (1993). *Cell Calcium*, **14**, 359.
3. Spencer, C. I. and Berlin, J. R. (1995). *Pflugers Arch.*, **430**, 579.
4. Scheenen, W., Makings, L. R., Gross, L. R., Pozzan, T., and Tsien, R. Y. (1996). *Chem. Biol.*, **3**, 765.
5. Pawley, J. (1995). *Handbook of biological confocal microscopy*. Plenum Press, New York.
6. Lipp, P. and Niggli, E. (1993). *Biophys. J.*, **65**, 2272.
7. Bootman, M., Berridge, M. J., Niggli, E., and Lipp, P. (1997). *J. Physiol.*, **499**, 307.
8. Yao, Y., Choi, J., and Parker, I. (1995). *J. Physiol. London*, **482**, 533.
9. Cheng, H., Lederer, W. J., and Cannell, M. B. (1993). *Science*, **262**, 740.
10. Lipp, P. and Niggli, E. (1994). *Circ. Res.*, **74**, 979.
11. Lopez-Lopez, J., Shacklock, P., Balke, C., and Wier, W. (1995). *Science*, **268**, 1042.
12. Koizumi, S., Bootman, M. D., Bobanovic, L. K., Schell, M. J., Berridge, M. J., and Lipp, P. (1999). *Neuron*, **22**, 125.
13. Niggli, E., Piston, D. W., Kirby, M. S., Cheng, H., Sandison, D. R., Webb, W. W., *et al.* (1994). *Am. J. Physiol.*, **266**, C303.
14. Lipp, P. and Niggli, E. (1996). *J. Physiol.*, **492**, 31.
15. Lipp, P. and Niggli, E. (1998). *J. Physiol.*, **508**, 801.
16. Grinkiewicz, G., Poenie, M., and Tsien, R. Y. (1985). *J. Biol. Chem.*, **260**, 3440.

Chapter 3

Detecting and minimizing errors in calcium-probe measurements arising from transition metals and zinc

Vladislav Snitsarev
Internal Medicine, University of Iowa, Iowa City, IA 52242, USA.

Alan R. Kay
Department of Biological Sciences, University of Iowa, Iowa City, IA 52242, USA.

1 Introduction

Transition metal ions and zinc play a central role in cells, serving as active elements at the heart of catalytic sites or in determining the tertiary structure of proteins (1). Such metal ions are often in very tight association with their proteins and do not readily exchange with the aqueous phase. In contrast, fluctuations in the concentration of free Ca^{2+} and Mg^{2+}, and Ca^{2+}'s role as second messenger are well known, primarily through the agency of fluorescent probes developed by Roger Tsien (2). There is, however, a great deal of uncertainty as to whether the other biologically important metals (Cu, Co, Fe, Mn, Mo, and Zn^{2+}, that we will for ease of reference refer to as 'transition metals' although Zn^{2+} is not strictly speaking a transition metal) (3) exist as free ions in the cytoplasm. (Note: We will refer to metal ions that exhibit variable oxidation states without a superscript.) Because of the vast difference in the biological chemistry of these ions, it is likely that the answers will differ from element to element and also from cell to cell.

Free Cu and Fe ions are potentially hazardous as they catalyse the formation of the highly reactive hydroxyl radical from peroxide (4) and thus cells have good reason to keep their free concentrations in check. Cu ions are ferried from plasma membrane transporters by metallochaperones to their specific apoenzymes (5). These chaperones in combination with intracellular chelators (e.g. metallothioneins) hold the free Cu ion levels at vanishingly low concentrations (less than 10^{-18} M) (6). However, there are cases where free metal concentrations may reach appreciable levels. For example Zn^{2+} in certain glutamatergic synaptic vesicles in the mammalian forebrain, appears to be in free ionic form or only weakly ligated and this Zn^{2+} appears to be released during synaptic transmission (7). There are also indications that free Fe concentrations may become significant under certain conditions (8).

Purely electrostatic considerations are not very informative when considering the interaction of transition metals with anionic ligands, as the particulars of the atomic orbitals of the metal play a determing role in the strength of interaction between a metal ion and ligand. For most ligands the transition metals follow the empirical Irving–Williams stability series viz.; $Mn^{2+} < Fe^{2+} < Co^{2+} < Ni^{2+} < Cu^{2+} > Zn^{2+}$ and all have higher affinities than Ca^{2+} and Mg^{2+} (9). This series has relevance to the topic under discussion as if the free metal ion concentration becomes elevated it may compete with Ca^{2+} or Mg^{2+} bound to natural ligands or fluorometric probes. Moreover, a rise in concentration of an ion like Cu may lead to the displacement of other ions leading the insertion of an inappropriate metal into enzymes or regulatory proteins, with its consequent inactivation.

Unlike transition metals, Ca^{2+} and Mg^{2+} enter into strictly electrostatic interactions and do not form coordination bonds. It is also worth noting that Cu, Fe, and Mn ions are redox active, while Ca^{2+}, Mg^{2+}, and Zn^{2+} are not, leading to differences in the way in which these groups of metals are treated. We will not consider other metal ions that are expected to be at much lower concentrations for example, Mo, Ni, Cr, and Co.

In cell biology the spatio-temporal dynamics of intracellular ion concentrations are widely monitored through fluorometric probes that experience a shift in their quantum yield or in their spectral characteristics on binding metals. Variations of ion concentrations can usually be estimated if the concentrations lie within an order of magnitude above and below the dissociation constant (K_d) of the probe. Fluorimetric probes have been used to monitor changes in Ca^{2+} concentration in the cytoplasm (10), nuclei (11, 12), endoplasmic reticulum (13), mitochondria (14), endosomes (15), and in the extracellular space (16). All of these probes are susceptible to interference from transition metals. In this chapter we examine how to determine if such interference has come into play, how to minimize it, and possibly how to correct such problems.

It was recognized in the first paper on fura-2 that the probe also responds to metal ions other than Ca^{2+} (2). Generally Zn^{2+} behaves qualitatively like Ca^{2+} in that it increases the fluorescence and shifts fluorescent peaks of probes like fura-2, while, Cu, Fe, and Mn, quench fluorescence. The latter effect occurs because the unpaired electrons in the d-orbitals of these ions enhances the probability of non-radiative transitions from the excited state (17). These effects occur at far lower concentrations than for Ca^{2+} because of the higher affinity for the probe. For example, the affinity of fura-2 for Zn^{2+} is approximately 100-fold higher than that for Ca^{2+} (2).

There is growing evidence suggesting that transition metals may in particular cells, severely distort measurements with fluorescent Ca^{2+} probes. For example, the spectral characteristics of fura-2 in the cytoplasm of hepatocytes and pancreatic acinar cells gave indications of interacting with Fe^{2+} and Zn^{2+} in addition to Ca^{2+} (18). Furthermore, free pools of Zn^{2+} have been found in synaptic vesicles, secretory granules (19), and spermatozoa (20), while elevated Fe ion concentrations have been found in hepatocytes, pancreatic acinar cells, and erythroleukemia K562 cells (18, 21).

2 Competition between endogenous transition metals and Ca^{2+} for fluorescent probes

In this chapter we consider errors in Ca^{2+} measurements arising from the association of a fluorometric probe with transition metal ions. We will not consider errors that arise from the incomplete de-esterification of AM derivatives, cytoplasmic viscosity, photobleaching, trapping of the probe in intracellular compartments, interaction with intracellular proteins (22, 23), or chemical agents (24).

If we consider the interaction of a fluorescent probe with Ca^{2+}, Fe^{2+}, and Zn^{2+}, the equilibrium is determined by the following reactions and the concentration of the metal–probe complexes will be determined by the dissociation constant (K_d) of each reaction.

$$\begin{aligned} Ca^{2+} + probe &\rightleftharpoons Ca^{2+} \cdot probe \\ Zn^{2+} + probe &\rightleftharpoons Zn^{2+} \cdot probe \\ Fe^{2+} + probe &\rightleftharpoons Fe^{2+} \cdot probe \end{aligned} \qquad [1]$$

The steady state concentrations of all components in this reaction can be determined if all K_ds are known using a variety of programs (25–27). Furthermore, if the forward and backward rate constants are known then the full time-dependent differential equation that governs this system can be solved, and the rate at which the system approaches equilibrium can be calculated.

We will not consider Cu and Mn ions, because of their low concentrations, however, should there be cells with elevated concentrations of these ions, they can be treated in the same way as Fe ions. Mg^{2+} has a very low affinity for most fluorescent Ca^{2+} indicators and does not present a substantial problem for Ca^{2+} measurements, although some fluorescent Ca^{2+} probes with low affinity to Ca^{2+} like mag-fura-2 (28) can be used to measure concentrations of Mg^{2+} in the low millimolar range (29, 30, and Chapter 6).

A fraction of intracellular iron (mainly Fe^{2+}) is associated with low molecular weight, low affinity chelators such as citrate, phosphate, and ATP, as well as proteins (31). Chelatable pools of iron (4–10 μM) have been found in the cytoplasm of some cells (18, 21) through its interaction with fluorescent probes.

In what follows we will show that there are diagnostic tests to determine whether or not transition metals might be interfering with fluorometric determinations of Ca^{2+} or Mg^{2+} concentrations.

3 Membrane-permeant transition metal chelators

In order to demonstrate that transition metals are interfering with fluorometric measurements it is necessary to employ a chelator that has a high affinity for these metals, yet not for Ca^{2+} and Mg^{2+} and crosses the plasma membrane (32). Examples of such chelators are shown in *Table 1*. Extreme caution needs to be exercised in using these agents, because they are likely to strip metals from a wide range of proteins, including Zn^{2+}-finger proteins, and adversely affect cellular function (33–35). We therefore recommend that these membrane-permeant chelators only be used for short-term exposures.

Table 1 Membrane-permeant transition metal chelators

Abbreviation	Structure	Chemical	Properties	Reference
DEDTC		Diethyldithiocarbamic	Water soluble	36
2,2′-DPD		acid 2,2′-dipyridyl	Sparingly soluble in water Dissolve the stock in alcohol or DMSO	21
TPEN		*N,N,N′,N′*-tetrakis(2-pyridylmethyl)ethylenediamine	Sparingly soluble in water Dissolve the stock in alcohol or DMSO	32, 18

We have found TPEN to be most useful in single cells and DEDTC in slices. It is worth noting that millimolar concentrations of TPEN have also been used to chelate the high concentrations of Ca^{2+} found in the endoplasmic reticulum (13, and Chapter 6).

If a membrane-permeant chelator is to be used with tissue slices it is important to ensure that the chelator penetrates into the depths of the slice. In hippocampal slices free Zn^{2+} in the synaptic vesicles can be visualized using the Zn^{2+}-specific fluorometric probe Zinquin (37). Application of a membrane-permeant metal chelator like DEDTC chelates the metal and prevents development of the metal-dependent fluorescence (38) and can thus serve as a natural assay for testing the penetration of metal chelators into tissue and its membrane permeance. We have found that DEDTC appears to penetrate into brain slices more effectively than TPEN.

4 Determining if fluorometric measurements of Ca^{2+} have been perturbed by transition metal ions

There are two broad classes of fluorometric ion probes, ratiometric (e.g. fura-2, mag-fura-2) and non-ratiometric probes (e.g. fluo-3, rhod-2) (39). The former exhibit shifts in their excitation or emission spectra on binding metals, and this shift can be used to make estimates of ion concentrations that are independent of path length and dye concentration. The non-ratiometric probes only exhibit changes in the intensity of fluorescence on binding metals, sometimes with very

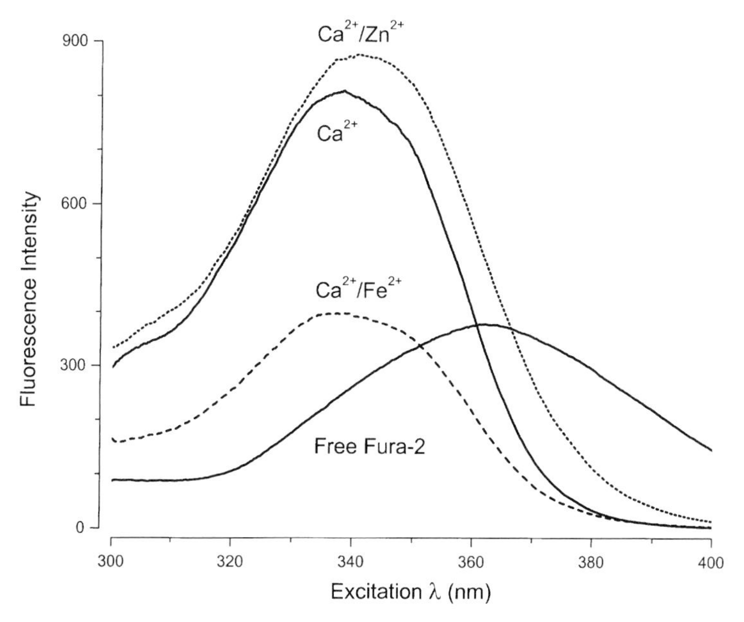

Figure 1 Perturbation of the fura-2 spectrum by Zn^{2+} and Fe^{2+}. Excitation spectra (emission 510 nm) of free fura-2 (1 μM), fura-2 saturated with Ca^{2+} (100 μM $CaCl_2$), fura-2 saturated with a mixture of Ca^{2+} (100 μM $CaCl_2$) and Fe^{2+} (20 μM $FeSO_4$) ions (Ca^{2+}/Fe^{2+}), and fura-2 saturated with a mixture of Ca^{2+} (100 μM $CaCl_2$) and Zn^{2+} (1 μM $ZnSO_4$) ions (Ca^{2+}/Zn^{2+}). The solution used for the Fe^{2+} spectrum was bubbled with nitrogen. All spectra were measured at 25°C in a cytosol-like solution containing 140 mM KCl, 20 mM NaCl, 10 mM MOPS pH 7.0 (NaOH) with contaminating transition and heavy metals removed according to *Protocol 2*.

small shifts of the spectra. The differences in the responses of these two classes of probes have important implications for their abilities to detect ions other than Ca^{2+} or Mg^{2+}.

Metals with incomplete d-orbitals (e.g. Cu, Fe, Mn, Mo, and Ni) have different effects on fluorometric probes than those with filled (Zn^{2+} and Cd^{2+}) or empty d-orbitals (La^{3+}). The former quench fluorescence (see *Figures 1* and *2B*), while the latter may increase or decrease the fluorescence relative to Ca^{2+}. In the case of ratiometric probes this last class of metals may lead to spectral shifts that are quite different from those induced by Ca^{2+} (see *Figures 1* and *2A*). The difference in the excitation spectra of fura-2 with Zn^{2+} bound and that with Ca^{2+} are shown in *Figure 2A*. Titrating the fluorometric probe saturated with Ca^{2+} with a metal ion allows one to estimate the relative affinity of the metal ion for the probe in the following way (*Figures 2C* and *D*).

The mass-action expressions for Ca^{2+} and Zn^{2+} are respectively:

$$K_{Ca} = \frac{[Ca^{2+}][Fura]}{[Ca \cdot Fura]} \text{ and } K_{Zn} = \frac{[Zn^{2+}][Fura]}{[Zn \cdot Fura]} \text{ therefore } \frac{K_{Ca}}{K_{Zn}} = \frac{[Ca^{2+}][Zn \cdot Fura]}{[Zn^{2+}][Ca \cdot Fura]}.$$

If fura-2 (1 μM) is saturated with Ca^{2+} (100 μM) and titrated with Zn^{2+}, free fura-2 concentration will be negligible, then when the concentration of Zn^{2+} at

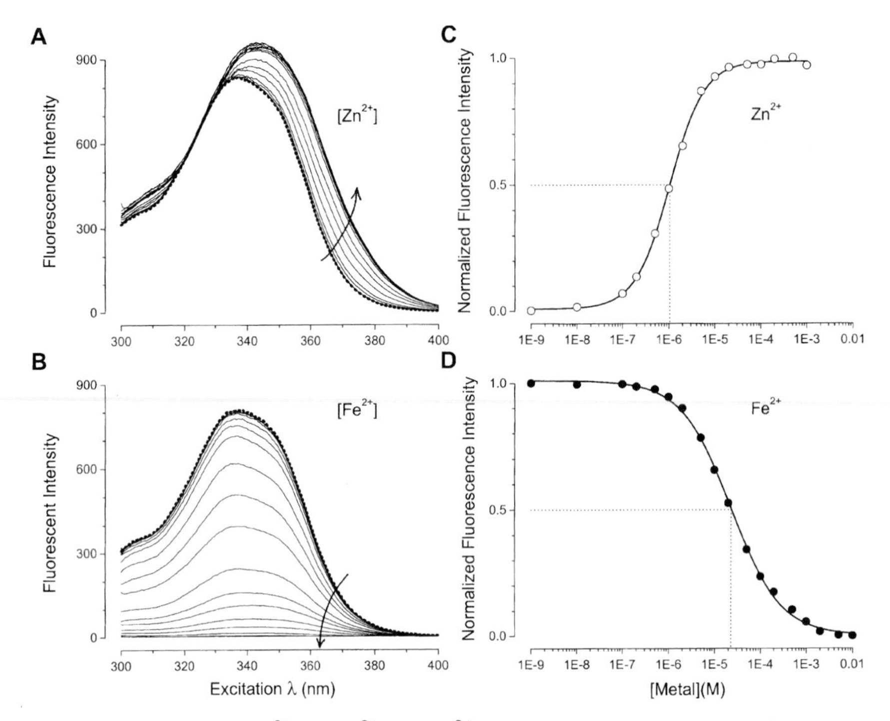

Figure 2 Competition of Zn^{2+} and Fe^{2+} with Ca^{2+} for fura-2. Excitation spectra (emission 510 nm) of fura-2 (1 μM) saturated with Ca^{2+} at different Zn^{2+} (A) and Fe^{2+} (B) concentrations (dotted line represents the spectrum in the absence of the transition metal; arrows indicate the direction of change with increasing metal concentrations). The normalized fluorescence intensity of these spectra (excitation 360 nm, emission 510 nm) was plotted as a function of Zn^{2+} (C) and Fe^{2+} (D) concentrations. Solutions and conditions were the same as those in *Figure 1*.

which the fluorescence is half-maximal ($Zn_{1/2}$), half of the fura-2 has Zn^{2+} bound. This means that, $[\boldsymbol{Ca \cdot Fura}] = [\boldsymbol{Zn \cdot Fura}]$, $[Ca^{2+}] \approx 100\ \mu M$, $[Zn^{2+}] \approx Zn_{1/2} - 0.5\ \mu M$, and

$$\frac{K_{Ca}}{K_{Zn}} \approx \frac{[Ca^{2+}]}{[Zn^{2+}]} \approx \frac{100\ \mu M}{Zn_{1/2} - 0.5\ \mu M}.$$

In these experiments the $K_{Ca}/K_{Zn} \approx 180$ (*Figure 2C*), $K_{Ca}/K_{Fe} \approx 4.5$ (*Figure 2D*), $K_{Ca}/K_{Cu} > 2000$, and $K_{Ca}/K_{Mn} \approx 34$ (not shown). In the case of Cu^{2+}, interaction is not 1:1, and the result is therefore only approximate.

Even the most highly purified salts will have some contaminating metals. For example the purest commercially available NaCl may have as much as 0.0001% of Fe, Zn, Pb, Cd, or other metals, that corresponds to approximately 100 nM in 100 mM NaCl solution. The other less predictable source of contamination is the dust and leaching labware. We recommend to avoid using metal labware (use plasticware or glassware instead) and wash carefully all labware that comes in contact with solutions with the highest purity deionized distilled water. The precise level of contamination (in ppb) can be determined by ICP-MS (inductively coupled plasma mass spectrometry), that is available at many universities, water analysis plants, or from a commercial provider. To determine if an artificial saline is contaminated with metal ions able to affect fluorometric Ca^{2+} measurements, we recommend using fura-2 to estimate the degree of contamination (see *Protocol 1*).

Protocol 1

Detecting contaminating heavy metals in solutions

Equipment and reagents

- Scanning spectrofluorometer
- Physiological solution to be tested
- Fura-2 K^+ salt
- 0.1 M EGTA pH 7.0 (with NaOH or KOH)
- 0.1 M TPEN in ethanol

Method

1. Make 6 ml of the Ca^{2+}-free saline to be tested for transition metals, add fura-2 (1 μM), and divide into two equal samples.
2. Measure the excitation spectrum from 300 nm to 400 nm (emission 510 nm) of sample 1.
3. Add EGTA (1 mM) to sample 1 and measure as in step 2 the excitation spectrum of free fura-2.
4. Add TPEN (10 μM) and $CaCl_2$ (100 μM) to sample 2 and measure as in step 2 the excitation spectrum of fura-2 saturated with Ca^{2+}.
5. Plot the three on the same axes. The spectrum of free fura-2 (measured in step 3) should cross the spectrum of fura-2 saturated with Ca^{2+} (measured in step 4) at 360 nm, which is the isosbestic point. If the spectrum measured in step 2 crosses the spectrum of free fura-2 to the left of the isosbestic point, fura-2 in the physiological solution has a quenching metal bound (see 'Ca^{2+}/Fe^{2+}' in *Figure 1*). If the spectrum measured in step 2 crosses the spectrum of free fura-2 to the right of the isosbestic point, fura-2 probably has Zn^{2+} bound (see 'Ca^{2+}/Zn^{2+}' in *Figure 1*). If the spectrum measured in step 2 crosses the spectrum of free fura-2 close to the isosbestic point, the fluorescence of fura-2 in the physiological solution has no significant contamination from transition and heavy metals.
6. If the contamination is a problem there are two ways to clean the solution:
   (a) Use weakly basic anion exchange resin (see *Protocol 2*).
   (b) Addition of EDTA and $CaCl_2$ in equimolar concentrations (in the range 10–100 μM) will chelate contaminating transition and heavy metals and will not affect free Ca^{2+} and Mg^{2+} concentrations.[a]

[a] EDTA dissociation constants: $Mg^{2+} = 10^{-8.9}$ M, $Ca^{2+} = 10^{-10.7}$ M, $Zn^{2+} = 10^{-16.5}$ M, $Fe^{2+} = 10^{-14.3}$ M, $Cu^{2+} = 10^{-18.8}$ M (9). In the presence of 1.5 mM Ca^{2+} and 1.2 mM Mg^{2+} that are concentrations usually used for extracellular salines, EDTA will hold contaminating transition and heavy metals below nanomolar concentrations.

In characterizing effects of expected endogenous cytoplasmic Zn^{2+} and Fe^{2+} on a fluorometric Ca^{2+} probe (as fura-2 in *Figure 2*) or cleaning a physiological solution of contaminating transition and heavy metals we recommend that the assay solution together with the probe or the physiological solution be purged of any contaminating transition metals (see *Protocol 2*, steps 6a or 6b).

Protocol 2

Removing transition metals from solutions

Equipment and reagents

- Scanning spectrofluorometer
- Column (11 mm × 40 mm)
- Nalgene Teflon FEP (fluorinated ethylene propylene) storage bottles; TEFLON (FEP)
- Diaion CR20 resin (Supelco)
- 1 M HCl
- Physiological solution that will be used making the measurements on the probe

Method

Weakly basic anion exchange resins prove effective in removing transition metals from solutions without perturbing the Ca^{2+} and Mg^{2+} concentrations. All stock solutions should be stored in pure Nalgene Teflon FEP bottles that have very low levels of leachable metals.

1. Carefully wash all labware that will come in contact with solutions with the highest purity deionized distilled water. Avoid using metal labware.
2. Regenerate Diaion CR20 resin by washing with 1 M HCl in a beaker three times.
3. Wash the resin with deionized distilled water in a beaker three times.
4. Fill the column with the resin to 50% of its height. Leave 1 cm of liquid on top of the resin bed.
5. Rinse the column with three bed volumes (~ 10 ml) of the physiological solution. Leave 1 cm of the solution on top of the resin bed.
6. (a) Fill the column with the physiological solution. Elute the column, collecting samples of the solution free of heavy and transition metals.
   (b) Fill the column with the physiological solution containing the fluorometric probe. Monitor the fluorescence of the eluent to detect the presence of the probe. Collect and pool all fractions containing the probe.

The response of non-ratiometric probes to Zn^{2+} depends on the nature of the probe. The following probes exhibit less fluorescence with Zn^{2+} bound than Ca^{2+} (both at 5 μM concentration): Calcium Green-1 (52%), Calcium Green-2 (54%), Calcium Green-5N (67%), Calcium Orange (88%), fluo-3 (61%), Oregon Green BAPTA-1 (55%), rhod-2 (53%). While the following probes experience a boosting of fluorescence with Zn^{2+} relative to that with Ca^{2+}: Calcium Crimson (104%), Magnesium Green (107%), Magnesium Orange (128%) (all data from ref. 40 with the exception of rhod-2, which we determined). For the first class of probes it is

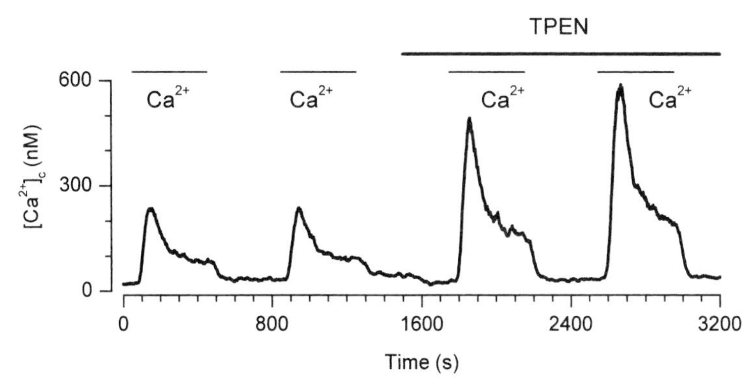

Figure 3 Relief of fluo-3 fluorescence by TPEN in the cytoplasm of cells. Calf pulmonary artery endothelial cells were loaded with fluo-3 AM (1 μM) for one hour at room temperature. The intracellular Ca^{2+} store was irreversibly emptied by thapsigargin in nominally Ca^{2+}-free saline. Readdition of extracellular Ca^{2+} (1.5 mM) causes $[Ca^{2+}]_c$ to rapidly increase and then decline to a plateau (54). Addition of TPEN (100 μM) increases the apparent intracellular Ca^{2+} changes (increases fluo-3 fluorescence; excitation 506 nm, emission 526 nm) evoked by external Ca^{2+} application, suggesting that endogenous transition metals mask the $[Ca^{2+}]_c$ responses in the first two Ca^{2+} applications.

not possible to distinguish the effect of Zn^{2+} from quenching metals using TPEN. For the second class of probes addition of TPEN would lead to a reduction of fluorescence if intracellular Zn^{2+} concentration is elevated.

Metal ions that quench fluorometric probes, reduce the effective concentration of the probe. If a ratiometric probe like fura-2 or mag-fura-2 is used to follow changes of Ca^{2+} concentrations, quenching transition metals should not lead to artefacts in these measurements, so long as the quenching metal concentrations remain stationary. However, in cases where non-ratiometric probes are used to quantify Ca^{2+} the presence of contaminating metals can significantly distort the estimates (41). An example of the sort of distortion induced by metals in cells is shown in *Figure 3*.

Application of a membrane-permeant transition metal chelator like TPEN and fura-2 as a metal probe can be used to test if intracellular transition metals might perturb determinations of intracellular Ca^{2+} (*Protocol 3*). If the intracellular concentration of a quenching metal is elevated then the application of TPEN will relieve the quenching of the fluorophore, but will not lead to a change in the ratio (see *Figure 4*). However, elevations of the free intracellular Zn^{2+} concentration may lead to significant distortions in the estimates of intracellular Ca^{2+} concentrations (11). Estimation of the ratio in the presence of a membrane-permeant transition metal chelator (*Table 1*) should correct for such errors. Experiments with non-ratioable fluorescent Ca^{2+} probes should be repeated in control with salines supplemented with one of the chelators (*Table 1*).

It might be thought that overloading a cell with a Ca^{2+} probe may be an effective strategy for minimizing the impact of endogenous transition metals.

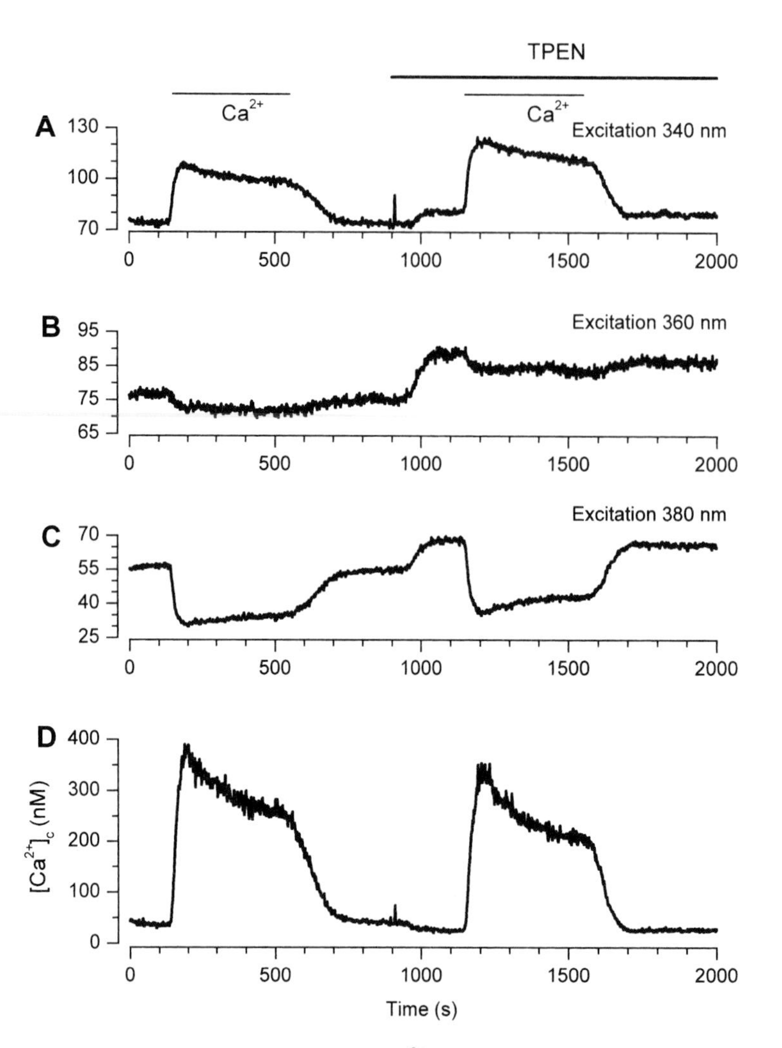

Figure 4 Ratiometric measurements of $[Ca^{2+}]_c$ are not affected by intracellular transition metals. Calf pulmonary artery endothelial cells were loaded with fura-2 AM (1 μM) for one hour at room temperature. The intracellular Ca^{2+} store was irreversibly emptied by thapsigargin in nominally Ca^{2+}-free saline. Readdition of extracellular Ca^{2+} (1.5 mM) caused $[Ca^{2+}]_c$ to rapidly increase and then decline to a plateau (54). Addition of TPEN (100 μM) increased fura-2 fluorescence at all excitation wavelengths, implicating a quenching metal ion. The ratiometrically derived overshoot in $[Ca^{2+}]_c$ is not affected by the presence of TPEN.

However, this may severely perturb intracellular Ca^{2+} levels (42) and may strip metals from intracellular proteins.

Although in the context of measuring Ca^{2+} the interference from transition metals is a nuisance, this property of fluorimetric probes can be used to detect transition metals. For example a number of groups have used fura-2 and mag-fura-2 to estimate intracellular Zn^{2+} concentrations (18, 43–48). In addition fura-2 has also been used to measure Cd^{2+} (49) and Pb^{2+} (50). The only chemical probe that can unequivocally distinguish between different metal ions in live cells is the NMR probe 5F-BAPTA, as the chemical shifts differ uniquely according to the type of metal bound (51–53).

Protocol 3

To determine if fura-2 measurements in cells are contaminated by transition metals

Equipment and reagents

- Scanning spectrofluorometer
- Fura-2 AM
- 0.1 M TPEN in ethanol

Method

1. Load cells with fura-2 AM.
2. Measure the excitation spectrum from 300 nm to 400 nm (emission, 510 nm).
3. Add TPEN (100 μM).
4. Measure the excitation spectrum as in step 2.
5. Plot the spectra measured in steps 2 and 4 on the same axes.
6. If the peak of the spectrum in step 4 is increased relative to that in step 2, then it is likely that fura-2 has some bound quenching metal, most likely Fe (possibly Cu or Mn). However, if the spectrum is reduced in amplitude it is likely that fura-2 has some bound Zn^{2+}. In this latter case ratiometric Ca^{2+} measurements should be made in the presence of one of the chelators listed in *Table 1*. In both cases one of the chelators should be used if Ca^{2+} measurements are made with non-ratioable fluorescent Ca^{2+} probe. If there is no change in fluorescent intensity at the isosbestic point (360 nm) on the addition of TPEN, this indicates that the measurements suffer no interference from transition metals.

Acknowledgements

This research was supported by NIH grant NS35243. We thank Dr Colin Taylor (Cambridge University) for his support of the initial stages of this work.

References

1. Lippard, S. J. and Berg, J. M. (1994). *Principles of bioinorganic chemistry*. University Science Books, Mill Valley, CA.
2. Grynkiewicz, G., Poenie, M., and Tsien, R. Y. (1985). *J. Biol. Chem.*, **260**, 3440.
3. Cotton, F. A., Wilkinson, G., Murillo, C. A., and Bochman, M. (1999). *Advanced inorganic chemistry*. John Wiley and Sons, Inc., New York.
4. Halliwell, B. and Gutteridge, J. M. C. (1999). *Free radicals in biology and medicine*. Oxford University Press, Oxford.

5. Malmstrom, B. G. and Leckner, J. (1998). *Curr. Opin. Chem. Biol.*, **2**, 286.
6. Rae, T. D., Schmidt, P. J., Pufahl, R. A., Culotta, V. C., and O'Halloran, T. V. (1999). *Science*, **284**, 805.
7. Frederickson, C. J. (1989). *Int. Rev. Neurobiol.*, **31**, 145.
8. Ferris, C. D., Jaffrey, S. R., Sawa, A., Takahashi, M., Brady, S. D., Barrow, R. K., *et al.* (1999). *Nature Cell Biol.*, **1**, 152.
9. Martell, A. E. and Hancock, R. D. (1996). *Metal complexes in aqueous solution*. Plenum Press, New York.
10. Berridge, M., Lipp, P., and Bootman, M. (1999). *Curr. Biol.*, **9**, R157.
11. Gerasimenko, O. V., Gerasimenko, J. V., Tepikin, A. V., and Petersen, O. H. (1996). *Pflügers Arch.*, **432**, 1.
12. Perez-Terzic, C., Stehno-Bittel, L., and Clapham, D. E. (1997). *Cell Calcium*, **21**, 275.
13. Hofer, A. M., Fasolato, C., and Pozzan, T. (1998). *J. Cell Biol.*, **140**, 325.
14. Hajnóczky, G., Robb-Gaspers, L. D., Seitz, M. B., and Thomas, A. P. (1995). *Cell*, **82**, 415.
15. Gerasimenko, J. V., Tepikin, A. V., Petersen, O. H., and Gerasimenko, O. V. (1998). *Curr. Biol.*, **8**, 1335.
16. Tepikin, A. V., Llopis, J., Snitsarev, V. A., Gallacher, D. V., and Petersen O. H. (1994). *Pflugers Arch.*, **428**, 664.
17. Lakowicz, J. R. (1983). *Principles of fluorescence spectroscopy*. Plenum Press, New York.
18. Snitsarev, V. A., McNulty, T. J., and Taylor, C. W. (1996). *Biophys. J.*, **71**, 1048.
19. Hutton, J. C., Penn, E. J., and Peshavaria, M. (1983). *Biochem. J.*, **210**, 297.
20. Andrews, J. C., Nolan, J. P., Hammerstedt, R. H., and Bavister, B. D. (1994). *Biol. Reprod.*, **51**, 1238.
21. Petrat, F., Rauen, U., and de Groot, H. (1999). *Hepatology*, **29**, 1171.
22. Roe, M. W., Lemasters, J. J., and Herman, B. (1990). *Cell Calcium*, **11**, 63.
23. Morris, S. A., Correa, V., Cardy, T. J., O'Beirne, G., and Taylor, C. W. (1999). *Cell Calcium*, **25**, 137.
24. Muschol, M., Dasgupta, B. R., and Salzberg, B. M. (1999). *Biophys. J.*, **77**, 577.
25. Parker, D. R., Norvell, W. A., and Chaney, R. L. (1994). In *Chemical equilibrium reaction models* (ed. R. H. Loeppert, S. Goldberg, and A. P. Schwab),Vol. XX. American Society of Agronomy, Madison, WI.
26. Fabiato, A. (1988). In *Methods in enzymology* (ed. S. Fleischer and B. Fleischer), Vol. 157, p. 378. Academic Press, San Diego, CA.
27. Schoenmakers, T. J. M., Visser, G. J., Flik, G., and Theuvenet, A. P. R. (1992). *BioTechniques*, **12**, 870.
28. Raju, B., Murphy, E., Levy, L. A., Hall, R. D., and London, R. E. (1989). *Am. J. Physiol.*, **256**, C540.
29. Stout, A. K., Li-Smerin, Y., Johnson, J. W., and Reynolds, I. J. (1996). *J. Physiol. (Lond).*, **492**, 641.
30. Brocard, J. B., Rajdev, S., and Reynolds, I. J. (1993). *Neuron*, **11**, 751.
31. Williams, R. J. P. (1982). *FEBS Lett.*, **140**, 3.
32. Arslan, P., Di Virgilio, F., Beltrame, M., Tsien, R. Y., and Pozzan, T. (1985). *J. Biol. Chem.*, **260**, 2719.
33. Lawrence, Y., Ozil, J. P., and Swann, K. (1998). *Biochem. J.*, **335**, 335.
34. Shumaker, D. K., Vann, L. R., Goldberg, M. W., Allen, T. D., and Wilson, K. L. (1998). *Cell Calcium*, **23**, 151.
35. Ahn, Y. H., Kim, Y. H., Hong, S. H., and Koh, J. Y. (1998). *Exp. Neurol.*, **154**, 47.
36. Danscher, G., Shipley, M. T., and Andersen, P. (1975). *Brain. Res.*, **85**, 522.
37. Zalewski, P. D., Forbes, I. J., and Betts, W. H. (1993). *Biochem. J.*, **296**, 403.
38. Budde, T., Minta, A., White, J. A., and Kay, A. R. (1997). *Neuroscience*, **79**, 347.
39. Minta, A., Kao, J. P. Y., and Tsien, R. Y. (1989). *J. Biol. Chem.*, **264**, 8171.

40. Haugland, R. P. (1996). *Handbook of fluorescent probes and research chemicals*. Molecular Probes, Eugene, OR.
41. Berman, M. C. (1999). *Biochim. Biophys. Acta*, **1418**, 48.
42. Neher, E. and Augustine, G. J. (1992). *J. Physiol. (Lond).*, **450**, 273.
43. Vega, M. T., Villalobos, C., Garrido, B., Gandia, L., Bulbena, O., Garcia-Sancho, J., *et al.* (1994). *Pflügers Arch.*, **429**, 231.
44. Hechtenberg, S. and Beyersmann, D. (1993). *Biochem. J.*, **289**, 757.
45. Simons, T. J. B. (1993). *J. Biochem. Biophys. Methods*, **27**, 25.
46. Atar, D., Peter, H. B., Appel, M. M., Gao, W. D., and Marban, E. (1995). *J. Biol. Chem.*, **270**, 2437.
47. Sensi, S. L., Canzoniero, L. M., Yu, S. P., Ying, H. S., Koh, J. Y., Kerchner, G. A., *et al.* (1997). *J. Neurosci.*, **17**, 9554.
48. Cheng, C. and Reynolds, I. J. (1998). *J. Neurochem.*, **71**, 2401.
49. Usai, C., Barberis, A., Moccagatta, L., and Marchetti, C. (1999). *J. Neurochem.*, **72**, 2154.
50. Schirrmacher, K., Wiemann, M., Bingmann, D., and Busselberg, D. (1998). *Calcif. Tissue Int.*, **63**, 134.
51. Smith, G. A., Hesketh, R. T., Metcalfe, J. C., Feeney, J., and Morris, P. G. (1983). *Proc. Natl. Acad. Sci. USA*, **80**, 7178.
52. Benters, J., Flögel, U., Schäfer, T., Leibfritz, D., Hechtenberg, S., and Beyersmann, D. (1997). *Biochem. J.*, **322**, 793.
53. Badar-Goffer, R., Morris, P., Thatcher, N., and Bachelard, H. (1994). *J. Neurochem.*, **62**, 2488.
54. Snitsarev, V. A. and Taylor, C. W. (1999). *Cell Calcium*, **25**, 409.

Chapter 4

Targeting of bioluminescent probes and calcium measurements in different subcellular compartments

Paulo J. Magalhães, Paolo Pinton, Luisa Filippin, and Tullio Pozzan
University of Padua, Department of Biomedical Sciences, Via G. Colombo 3, 35121 Padua, Italy.

Marisa Brini
University of Padua, Department of Biochemistry, Via G. Colombo 3, 35121 Padua, Italy.

Anna Chiesa and Rosario Rizzuto
University of Ferrara, Department of Experimental and Diagnostic Medicine, Section of General Pathology, Via L. Borsari 46, 44100 Ferrara, Italy.

1 Introduction

The medusa *Aequorea victoria* produces a 22 kDa protein that has played a major role in the study of calcium signalling. Aequorin, in its active form, is made up of a polypeptide moiety covalently bound to a prosthetic group, coelenterazine. Upon binding of Ca^{2+} to its high affinity sites, aequorin undergoes an irreversible reaction that causes the release of the prosthetic group and the emission of a photon (1–5).

Due to its Ca^{2+}-triggered light emission and the reliable calibration procedures, aequorin was a widely employed probe for the study of intracellular Ca^{2+} during the late 60s and most of the 70s. Milestone concepts of Ca^{2+} signalling—such as the oscillatory pattern of the agonist-dependent rises of cytoplasmic Ca^{2+} concentration—were established using this probe. In spite of its usefulness, aequorin presented a serious drawback. Since it is membrane impermeant, the protein had to be microinjected, limiting its use to large sessile cells. This limitation caused the gradual substitution of this probe with trappable fluorescent indicators (6–8).

In 1985 the cDNA for aequorin was isolated (9), and modern molecular biology techniques have endowed this probe with renewed potentialities. First, it is now possible to express aequorin within any cell type amenable to transfection, thus overcoming its original limitation. Secondly, it is possible to design chimeras of this protein directly targeted to specific subcellular locations. In the current chapter we will focus on this last theme, discussing the various strategies adopted to construct a fine-structure map of subcellular Ca^{2+} spatio-temporal dynamics.

2 Background principles

The active form of aequorin is made up of an apoprotein moiety covalently bound to coelenterazine. The expression of aequorin's cDNA yields the polypeptide, to which must be added the prosthetic group. Coelenterazine is added to the culture medium of transfected cells, it is allowed to diffuse freely across the cell membrane, and once inside the cell spontaneously binds to the polypeptide, generating the active probe. This procedure is generally termed 'reconstitution' and will be described in detail below.

The Ca^{2+}-dependent photon emission is depicted in *Figure 1A*. Binding of Ca^{2+} ions to aequorin's high-affinity sites promotes an irreversible reaction, in which coelenterazine is converted to coelenteramide and light is emitted. *Figure 1B* shows the relationship between this light emission and $[Ca^{2+}]$. The ratio between L, the rate of light emission at a given $[Ca^{2+}]$, and L_{max}, the maximal rate of the light emission at saturating $[Ca^{2+}]$, defines the fractional rate of aequorin consumption. For Ca^{2+} concentrations between 10^{-7} and 10^{-5} M, light from aequorin is proportional to the 2.5–3.0 power of calcium. When observed on a double log plot, values in this range fall conveniently on a steep straight line. The strength of this response and its wide dynamic range make aequorin particularly well suited for a variety of physiological studies. Furthermore, as will be discussed later in this chapter, changes in specific parameters enable this probe to be used in regions of much higher Ca^{2+} concentrations.

Photons emitted during the experimental procedure can be recorded and analysed using relatively inexpensive equipment assembled in-house; interested readers are referred to ref. 10 for excellent detailed technical descriptions. The schematic representation of such an arrangement is depicted in *Figure 2*. Essentially, coverslips containing the cells to be analysed are placed in a perfusion chamber, in close proximity to a photomultiplier. The perfusion chamber is maintained at 37 °C, which enables the cells not only to be kept under normal physiological conditions, but also to be challenged with any chosen substance during the experiment. The complete assemblage is housed in a light-tight chamber refrigerated to 4 °C, to reduce background interference. The photomultiplier signal is captured by a photon-counting board installed in an IBM PC-compatible computer. Cells are lysed at the end of every experiment, in the presence of high $[Ca^{2+}]$, in order to discharge all the remaining aequorin, which permits the calculation of L_{max}; the program can then convert the signals acquired during the experiment into values of Ca^{2+} concentration.

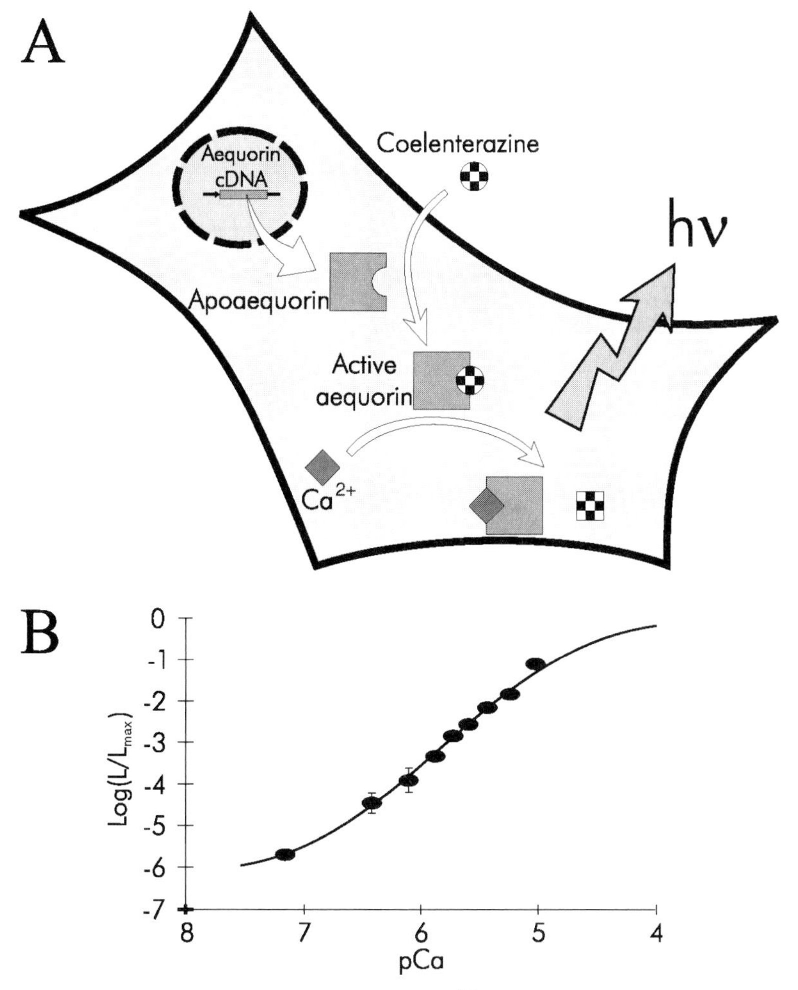

Figure 1 (A) Schematic representation of the Ca^{2+}-dependent photon emission in cells transfected with aequorin cDNA. The functional probe is generated from the apoaequorin moiety (synthesized by the cell) and coelenterazine (imported from the surrounding medium). Binding of calcium ions disrupts the binding of the prosthetic group with concomitant light emission. (B) Relationship between calcium concentration and light emission rate by active aequorin (L and L_{max} are, respectively, the instant and maximal rates of light emission). Note the direct relationship at calcium concentrations between 10^{-5} and 10^{-7} M.

3 Designing chimeric aequorin variants

Wild-type aequorin expressed in transfected cells is restricted to the cytosolic fraction, and is thus adequate for the measurement of $[Ca^{2+}]$ in this compartment. However, it is possible to modify the intracellular destination of the photoprotein by adding specific targeting signals to its primary sequence. Thus, an exciting new field—opened by the isolation of aequorin's cDNA—is that of developing Ca^{2+} probes specifically localized to different subcellular compart-

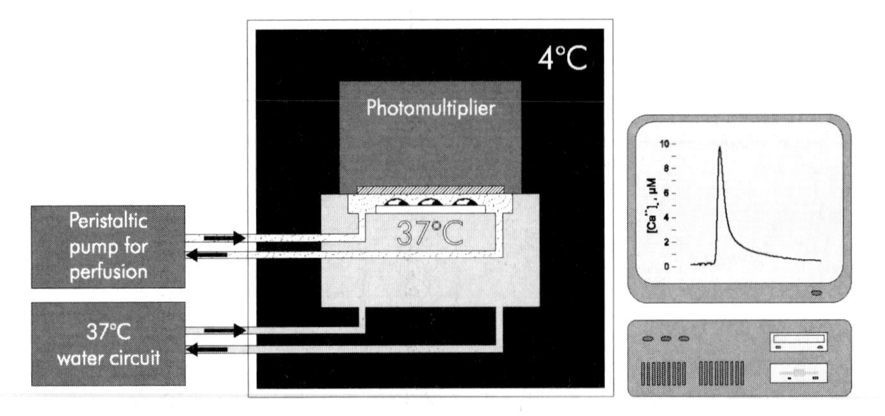

Figure 2 Schematic representation of a custom-built luminometer. Cells expressing functional aequorin probe are incubated in a perfusion chamber, at 37 °C, in close proximity to a photon-counting tube. The complete assemblage is kept at 4 °C, in the dark, to minimize extraneous signals. Acquisition of the data and subsequent calculations to transform light emission into $[Ca^{2+}]$ are performed by a dedicated computer.

ments. In a later section, we will broach the general requisites for specific targeting and describe the strategies used in the case of aequorin.

The first step in calcium measurements, common to any of the engineered variants of aequorin, is the introduction of the encoding cDNA into the cell type of interest. The most widely used method in our laboratory is that of calcium phosphate (11); it is described in detail in *Protocol 1*.

Protocol 1

Calcium phosphate transfection of HeLa cells

Equipment and reagents

- 13 mm diameter coverslip
- 24-well plate
- Microcentrifuge tubes
- Trypsin: 0.1% trypsin, 0.1% EDTA, in PBS
- PBS: 10 mM Na_2HPO_4, 1 mM KH_2PO_4, 137 mM NaCl, 2.7 mM KCl pH 7.4
- Cell culture medium: DMEM, 2 mM glutamine, supplemented with antibiotics (streptomycin and penicillin), and 10% FCS
- Sterile TE: 10 mM Tris–HCl, 1 mM EDTA pH 7.6
- Sterile 2.5 M $CaCl_2$
- 2 × HBS: 280 mM NaCl, 50 mM Hepes, 1.5 mM Na_2HPO_4, pH *exactly* to 7.12, at RT, with 0.5 M NaOH; sterilize through a 0.22 μm filter
- 3 M sodium acetate pH 5.2
- Ethanol

Method

The following protocol refers to the standard transfection of a single coverslip; it can be scaled up according to the experimental requirements. See the text for exceptions with specific modifications.

Protocol 1 continued

Day 1

1 Trypsinize the cells and seed them at 30–50% confluence on a 13 mm diameter coverslip, placed in a standard 24-well plate, with 1 ml of culture medium.

Day 2

1 Transfer 4 μg of plasmid DNA (transfection-quality DNA, purified either by CsCl gradient or by anion exchange columns) into a clean 1.5 ml microcentrifuge tube, and ethanol precipitate the DNA (add 0.1 vol. of 3 M sodium acetate pH 5.2, 2.5 vol. of absolute ethanol, and mix well).

2 Collect the precipitated DNA by centrifugation at ≥ 15 000 g for 30 min, at RT; discard the supernatant.

3 Dissolve the precipitated DNA pellet in 45 μl of sterile TE and add 5 μl of 2.5 M $CaCl_2$.

4 Into a separate 1.5 ml microcentrifuge tube, pipette 50 μl of 2 × HBS.

5 In a sterile hood, slowly add—*dropwise!*—the contents of the first tube (DNA and $CaCl_2$) to the second (2 × HBS), while vortexing the latter.

6 The previous step is crucial for the transfection efficiency and should be carried out with the utmost care. The solution should become cloudy, indicative of a fine precipitate of DNA. Incubate the precipitate for 30 min at RT.

7 In the meantime, replace the culture medium of the cells to be transfected with fresh medium.

8 Add—slowly and dropwise—the cloudy DNA precipitate to the well containing the coverslip with cells.

9 Incubate overnight in standard cell culture conditions (37 °C, humidified atmosphere containing 5% CO_2).

Day 3

1 Wash the cells repeatedly with PBS, until all traces of Ca phosphate precipitate are removed—usually three to four times are sufficient.

2 Add 1 ml of normal medium and return the cells to standard cell culture conditions. In general, transfected cells are ready for aequorin measurements 24–36 h after the transfection.

Introduction of foreign DNA into cells is generally feasible using the calcium phosphate method. Certain cell types, however, are particularly resilient to this procedure. In such cases the solution has been to seek alternative methods for the introduction of cDNA into that particular cell type. Procedures which we have found to fit better specific cell types include electroporation, Boehringer Mannheim's FuGENE 6 Transfection Reagent, Gibco BRL's LipofectAMINE Reagent, and Bio-Rad's Biolistic microprojectile gene delivery system (commonly termed 'gene gun'). Since detailed protocols are supplied by the manufacturers of both kits and specific instrumentation, these will not be discussed here. In our experience, we have empirically determined that all cell types in which we have tried to express aequorin can be successfully transfected using one method or another.

Cells expressing any of the engineered variants of aequorin require coelenterazine for the active form to be generated. This is achieved by incubating the cells in the presence of the prosthetic group, as described in *Protocol 2*.

Protocol 2

Reconstitution of active aequorin

Equipment and reagents

- Coverslip
- Tissue culture plates
- Luminometer chamber
- KRB: 125 mM NaCl, 5 mM KCl, 1 mM Na_3PO_4, 1 mM $MgSO_4$, 5.5 mM glucose, 20 mM Hepes pH 7.4, at 37 °C
- Cell culture medium (see *Protocol 1*)
- 100 × coelenterazine solution: coelenterazine (Molecular Probes) is dissolved at a final concentration of 0.5 mM in pure methanol, aliquoted, and kept at -80 °C in the dark; coelenterazine is *very* sensitive to light

Method

The following is a standard procedure for general use; see the text for exceptions with specific modifications.

1 Remove the normal culture medium from the well containing the coverslip with transfected cells.

2 Add 1 ml of culture medium (with *only 1%* FCS), and supplement with coelenterazine to a final concentration of 5 μM.

3 Incubate for 2 h under standard cell culture conditions (see *Protocol 1*), in the dark.

4 Transfer the coverslip with the cells directly to the luminometer chamber and initiate perfusion with KRB.

It is particularly convenient that coelenterazine is capable of freely crossing all cell membranes. The engineered versions of aequorin, independent of their subcellular localization, are thus capable of being reconstituted into the active form of the probe. It should be noted, however, that during the reconstitution time (1–2 hours is generally sufficient for optimal results), active aequorin is being consumed due to the Ca^{2+} present. Whereas in compartments with low $[Ca^{2+}]$ this does not present a problem (high rate of reconstitution of a functional aequorin pool versus a negligible consumption), specific counteractions must be employed in compartments with high $[Ca^{2+}]$. Such measures are described in a later section.

Analyses of transfected cells harbouring a pool of functional aequorin are made in a luminometer (*Figure 2*). Our custom-built version essentially consists of a perfusion chamber (where the coverslip with cells is placed, and maintained in controlled conditions), on top of a hollow cylinder kept at a constant temperature of 37 °C by water flow. The cells are within a few millimetres of a low-noise phototube, and the complete assemblage is kept in a light-tight chamber, maintained at 4 °C. The acquisition is performed via custom-made programs with the aid of software provided by the manufacturer of the photon-counter (Thorn-EMI).

During each experiment the perfusion medium can be manipulated and the cells' responses recorded. An adjustable peristaltic pump used for perfusion enables a quick and easy control of the incubation conditions. On one hand, the flow can be maintained at a minimum for normal perfusion with unchanged parameters; however, challenging the cells with any desired agent can be achieved rapidly simply by introducing the appropriate solution into the system and temporarily increasing the perfusion flow rate.

Transformation of the raw luminescent signal into $[Ca^{2+}]$ is achieved with an algorithm that takes into account the instant rate of photon emission, the conditions of pH, ionic strength, $[Mg^{2+}]$, and the total amount of light that can be generated by the aequorin in the sample. It is for the latter that every experiment ends with the complete lysis of the cells, in a medium containing a high concentration of calcium—thus ensuring a complete discharge of all available reconstituted aequorin. The acquired and recorded raw data is then calibrated using a custom-made program. Alternatively, it can be exported and the calculations made in any non-specific data handling program, such as Microsoft's *Excel* or Microcal's *Origin*, as described in detail in ref. 12. Final traces reproduce the variability of $[Ca^{2+}]$ in response to different stimuli.

Figure 3A shows a raw data trace of luminescence variation in HeLa cells

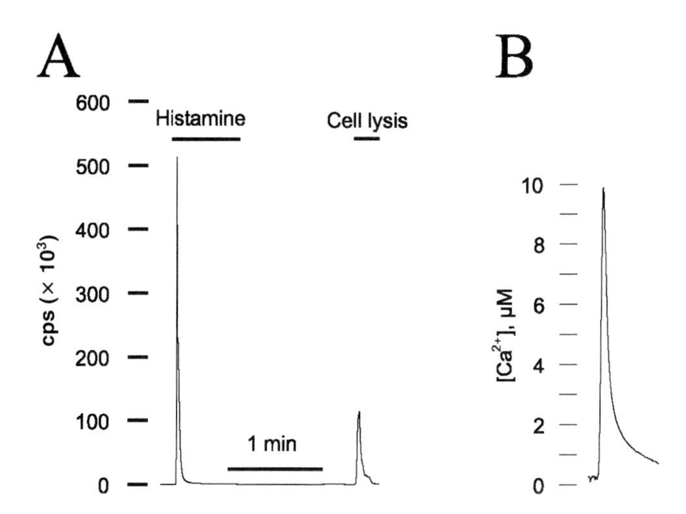

Figure 3 Transformation of photon counts into $[Ca^{2+}]$ values. Light emission (A) and calculated $[Ca^{2+}]$ values (B), from a monolayer of HeLa cells expressing aequorin targeted to the mitochondrial matrix. Where indicated, the cells were challenged with histamine (100 μM). At the end of the experiment the cells were lysed to estimate the total amount of photoprotein. cps, counts per second.

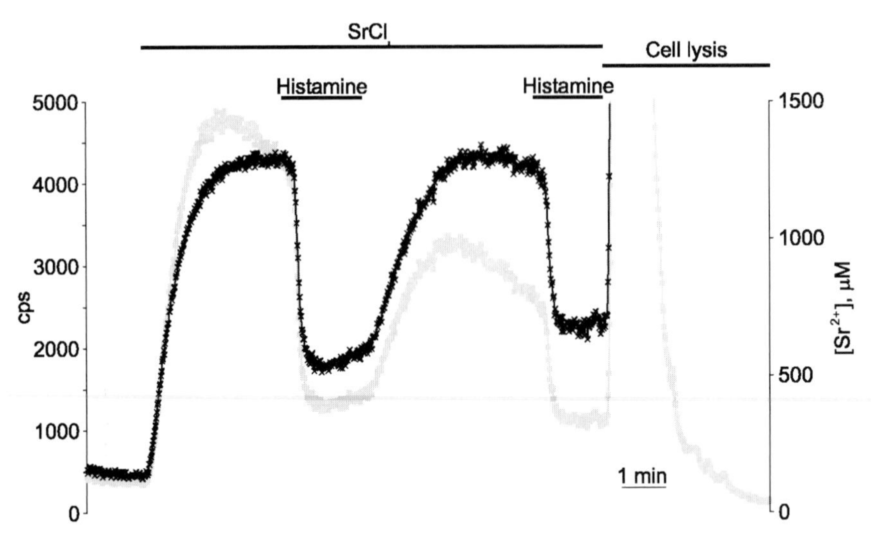

Figure 4 Relationship between photon counts and calculated concentrations. HeLa cells expressing aequorin targeted to the Golgi apparatus were initially emptied of Ca^{2+} (see *Protocol 3*), refilled with the surrogate cation Sr^{2+} (see text for details), and subsequently challenged with histamine before the final cell lysis. The trace in grey represents photon counts (scale at left; cps, counts per second), while the trace in black shows the calculated $[Sr^{2+}]$ (scale at right). Note that upon initial refilling of the Golgi apparatus with Sr^{2+}, the photon count rises to a peak and naturally begins to decrease—this decrease is due to the consumption of aequorin itself, and the calculated trace correctly shows that $[Sr^{2+}]$ reaches a plateau level that is naturally maintained. Addition of histamine causes the release of Sr^{2+} from this intracellular store, as shown clearly by both traces. Removal of histamine from the system enables the levels of Sr^{2+} in the Golgi apparatus to be re-established. Note, however, that the maximal photon count is now much lower (aequorin is continuously consumed during the experiment), while the $[Sr^{2+}]$ plateau is similar to that obtained previously, as expected. A second challenge with histamine produces an analogous Sr^{2+} release. The final cell lysis enables the measurement of the total amount of aequorin remaining in the system, providing the necessary information for the calibration procedure.

transiently expressing aequorin targeted to the mitochondrial matrix. Upon stimulation with histamine, the intensity of the signal rises several-fold. Note that the relationship between light emission and $[Ca^{2+}]$ is not linear, but exponential; i.e. for $[Ca^{2+}]$ between 10^{-5} and 10^{-7} M the light signal generated is proportional to the 2nd–3rd power of $[Ca^{2+}]$. *Figure 3B* shows the calculated $[Ca^{2+}]$ trace, based on the recorded data. The ease of calibration and the excellent signal-to-noise ratio enable this chemiluminescent probe to provide very clean measurements. *Figure 4* demonstrates the apparent discrepancy sometimes observed between the dynamics of the initial raw signal and the final calculated trace.

3.1 Bioluminescent probes for mitochondrial matrix and mitochondrial intermembrane space

With the exception of 13 polypeptides encoded in the mitochondrial genome, all other mitochondrial proteins are encoded in nuclear DNA, synthesized by

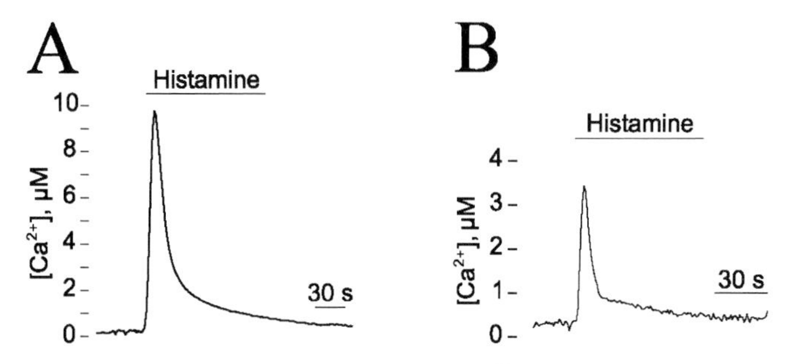

Figure 5 Agonist-dependent increases in Ca^{2+} concentrations in HeLa cells expressing aequorin targeted to the mitochondrial matrix (A) and the intermembrane space (B). Where indicated, the cells were challenged with 100 μM histamine.

cytosolic ribosomes, and imported into mitochondria. To ensure proper delivery to this organelle, proteins are generally made with a signal peptide, usually present at their N-terminus. Different proteins possess different signal peptides, with no obvious consensus sequence, which simply bear some common characteristics regarding their charge distribution (13).

Subunit VIII of human cytochrome *c* oxidase possesses a 25 amino acid signal peptide at its N-terminus, which is cleaved by a matrix protease upon import; it is presumed that the action of the protease is dependent on a motif present within the first few amino acids of the mature protein (14). We fused the initial 31 amino acids of this subunit, comprising the signal peptide and the first 6 amino acids of the mature protein, to the N-terminus of aequorin, thus achieving the correct delivery of functional aequorin to the mitochondria matrix (15).

For the delivery of aequorin to the mitochondrial intermembrane space (MIMS), we exploited the characteristics of another mitochondrial protein. Glycerol phosphate dehydrogenase (GPD) is an enzyme present in the mitochondrial inner membrane with a C-terminal domain protruding into the MIMS. To target aequorin to this space, we fused the photoprotein to the C-terminal portion of GPD, thus maintaining aequorin's C-terminus unaltered, since it is essential for its luminescent properties (16).

Measurements obtained in HeLa cells transiently expressing aequorin targeted to the mitochondrial matrix and the mitochondrial intermembrane space are shown in *Figure 5*.

3.2 Measurement of ER and SR calcium using targeted aequorin

Proteins naturally destined to the ER usually contain a double targeting signal. First, an hydrophobic sequence located at the protein's N-terminus directs its translation on membrane-bound ribosomes and its insertion into the ER. Second, a signal located at the protein's C-terminus prevents its escape from the ER; one of the best characterized of these latter signals is the tetrapeptide KDEL (17). In analogy to the mitochondrial cases, addition of such sequences to aequorin should ensure a correct localization. However, as mentioned above, modification of the C-terminus of aequorin severely impairs its chemiluminescent properties, so an alternative strategy had to be devised.

To ensure the correct delivery of aequorin to the ER we fused a fragment comprising the L, VDJ, and CH1 domains of to the N-terminus of aequorin. Under natural conditions, the mentioned domains (particularly CH1 that binds to BiP, an endogenous ER protein) enable the heavy chain of immunoglobulins to be retained in the ER until assembly with the light chain takes place (18).

The final chimera also harboured a point mutation at position 119 of wild-type aequorin (Asp → Ala), which reduces the affinity of the photoprotein to calcium (19); on this theme, see also the section on compartments with high calcium concentration.

For the SR-targeted chimera, the approach was straightforward, being based simply on the fusion of mutated aequorin to the C-terminus of calsequestrin, an endogenous SR protein (20).

Measurements obtained with recombinant aequorin targeted to the ER (in HeLa cells) and the SR (in primary cultures of rat skeletal muscle myotubes) (21) are shown in *Figures 6A* and *6B*.

3.3 Targeted aequorin as an instrument of calcium measurement in the Golgi apparatus

To measure variations in Ca^{2+} concentration within the Golgi apparatus, we grafted a fragment from sialyltransferase (ST) onto aequorin; ST is an endogenous

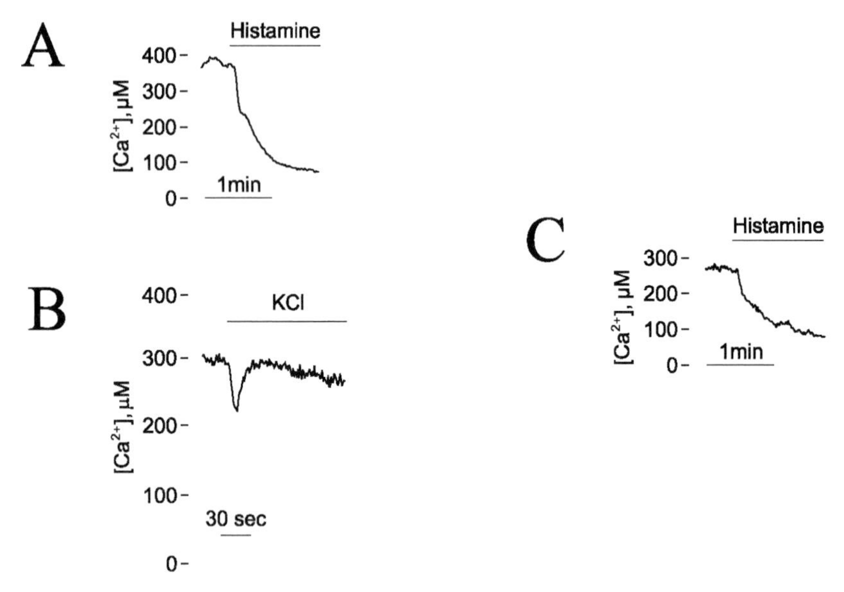

Figure 6 Calcium release from internal stores. HeLa cells expressing aequorin targeted to the ER (A), and Golgi apparatus (C) exhibit a marked drop in $[Ca^{2+}]$ in these internal stores when challenged with 100 μM histamine. In rat myotubes expressing aequorin targeted to the SR, a typical transient release of Ca^{2+} is observed upon stimulation with 130 mM KCl (B).

protein of the lumen of the *trans*-Golgi and *trans*-Golgi network which is kept in place due to a transmembrane domain. The chimeric polypeptide was obtained by fusing the cDNA encoding a 17 amino acid membrane-spanning domain from ST, with the cDNA encoding mutated (D119A) aequorin. The ST portion of the fusion protein effectively retains the protein within the Golgi. *Figure 6C* shows measurements obtained in HeLa cells transiently expressing aequorin targeted to the Golgi apparatus.

3.4 Cytoplasmic and plasma membrane targeted bioluminescent probes

Cells transfected with wild-type aequorin cDNA exhibit the photoprotein exclusively in the cytosolic fraction. No modifications are required, therefore, to measure [Ca^{2+}] in this compartment.

It is possible to target aequorin to specific regions of the cytosol, such as, for example, the subplasmalemmal space.Our first attempt to design a chimeric aequorin targeted to this domain was based on the fusion of an integral protein of the plasma membrane, the metabotropic glutamate receptor mGluR1. The fusion protein, while exhibiting the appropriate Ca^{2+}-dependent luminescence properties, did not show the expected subcellular destination. Indeed, it was largely retained within intracellular vesicular compartments (ER, Golgi, vesicles) and thus could not be employed as a subplasmalema Ca^{2+} probe. Conversely, the second chimera constructed was very efficient. SNAP-25 is a protein that is synthesized on free ribosomes and, after palmitoylation of specific cysteine residues, it is recruited to the inner surface of the plasma membrane. Fusion of this protein to the N-terminus of aequorin effectively directs the chimeric peptide to that region (22).

3.5 Nuclear targeting of aequorin

Proteins destined to the nucleus generally possess a specific signal, known as nuclear localization signal (NLS), within their primary sequence. Comparisons of these sequences has led to the consensus that the SV40 T antigen NLS is the prototypic NLS. In general, these signals share common characteristics (8–10 amino acids long, rich in charged amino acids such as lysine and arginine, with no preferential site within the protein), they are *not* removed following the localization, and can occur more than once in a given protein.

Our strategy was based on the targeting sequence of the rat glucocorticoid receptor (GR). This receptor possesses an NLS which is masked by the interaction of a heat-shock protein, and is therefore restricted to the cytosol. Upon binding of the hormone, the NLS is exposed and translocation into the nucleus occurs (23). A foreign protein fused to GR will remain in the cytoplasm until the cells are treated with glucocorticoids, when it will be transported into the nucleus. However, if only a fragment of GR—lacking the hormone-binding domain—is used, the NLS is permanently unmasked and the fusion protein is constitutively delivered to the nucleus.

Based on the above, we have constructed two variants of aequorin chimeras for the measurement of nuclear [Ca^{2+}]: one constitutively present in the nucleus (nuAEQ), while the other (cyt/nuAEQ) is mobilized only in the presence of glucocorticoids (24).

4 Use of aequorin in the presence of high Ca^{2+} concentrations

Although aequorin is perfectly suited for measuring [Ca^{2+}] between 0.5 and 10 μM, it is unsuitable for measurements in regions where [Ca^{2+}] is much higher (e.g. the lumen of the ER and SR, near Ca^{2+} channels, pumps, etc.). This limitation can be overcome by reducing the affinity between Ca^{2+} and the photoprotein. At least three methods can be envisaged, and all three have been used successfully:

(a) Mutation of one or more of the Ca^{2+}-binding sites (25).

(b) Use of surrogate cations, such as Sr^{2+}, which elicit a slower rate of photoprotein consumption than Ca^{2+} itself (19).

(c) Use of modified prosthetic groups, such as coelenterazine *n*, which decrease the affinity of aequorin for Ca^{2+} (26).

Specifically, regarding the first item, mutating position 119 from Asp to Ala affects one of aequorin's Ca^{2+}-binding sites. The protein thus mutated has a markedly impaired affinity for the cation, and is capable of measuring [Ca^{2+}] in the range of 10–100 μM. Regarding surrogates, the divalent cation Sr^{2+} has been shown to be a suitable Ca^{2+} substitute: it can cross Ca^{2+} channels, and is actively transported (although with lower affinity) by both plasma membrane and sarco-endoplasmic Ca^{2+} ATPases. It is noteworthy that changing these two parameters in parallel enables the measurements to stretch into the mM range.

Currently, in order to avoid possible discrepancies between Ca^{2+} and Sr^{2+}, and also to specifically monitor [Ca^{2+}] variations in compartments with high [Ca^{2+}], the use of Sr^{2+} is falling in disuse. This is made possible because of the availability of a low affinity coelenterazine analogue (coelenterazine *n*), which gives rise to an aequorin–coelenterazine functional probe with a very low rate of consumption. Using Ca^{2+}, this synthetic prosthetic group, and mutated aequorin, it is possible to monitor millimolar concentrations of Ca^{2+} for relatively long periods of time.

In practical terms, and as alluded to in earlier parts of this chapter, measuring [Ca^{2+}] in regions of high calcium concentration (such as ER, SR, and Golgi apparatus) involves one final countermeasure. Since the functional aequorin–coelenterazine probe is irreversibly destroyed by Ca^{2+} ions, care must be taken so that sufficient intact functional probe is available at the beginning of each experiment; in other words, the active probe being generated during the reconstitution period (see above) must not be consumed at a similar or higher rate, as would be the case in the presence of high [Ca^{2+}]. To ensure this crucial point, it is necessary to promote a drastic reduction of the lumenal Ca^{2+} of such compartments before the reconstitution process. This procedure is described in *Protocol 3*.

Protocol 3

Emptying high Ca^{2+} stores prior to aequorin reconstitution

Reagents

- 100 × coelenterazine *n* solution (Molecular Probes): prepared as normal coelenterazine—see *Protocol 2*
- 100 mM tBuBHQ solution in DMSO (store in aliquots at −20 °C)
- 100 mM histamine solution (store in aliquots at −20 °C)
- KRB (see *Protocol 2*)
- 500 mM EGTA solution pH 7.4

Method

When it is necessary to deplete $[Ca^{2+}]$, different molecular Ca^{2+} transport pathways can be used. The sections that follow describe the specific procedures for different intracellular compartments. After completion of the steps described, the cells are ready for standard luminescence readings.

A ER

1 Incubate the cells for 5 min with tBuBHQ (a SERCA inhibitor) and histamine (an IP_3-generating agonist), both at a final concentration of 100 μM, in KRB supplemented with 3 mM EGTA.

2 Wash the cells with KRB containing 100 μM EGTA, 5% bovine serum albumin (BSA), and 10 μM tBuBHQ.

3 Reconstitute aequorin by incubating the cells at 25 °C, for 1 h with KRB containing 100 μM EGTA, 10 μM tBuBHQ, and 5 μM coelenterazine.

4 Wash the cells extensively with KRB containing 2% BSA and 1 mM EGTA.

B SR

1 Incubate myotubes for 2 min with KRB containing 3 mM EGTA, 10 mM caffeine, and 30 μM tBuBHQ.

2 Reconstitute aequorin by incubating the cells at 4°C, for 1 h with KRB containing 100 μM EGTA and 5 μM coelenterazine.

3 Wash the cell monolayer extensively with KRB containing 2% BSA and 1 mM EGTA.

C Golgi apparatus

1 Wash the cells with KRB supplemented with 600 μM EGTA.

2 Reconstitute aequorin by incubating the cells at 4 °C, for 1 h with KRB containing coelenterazine (5 μM), the Ca^{2+} ionophore ionomycin (5 μM), and 600 μM EGTA.

3 Wash the cells extensively with KRB containing 2% BSA and 1 mM EGTA.

Note that this procedure can also be used for the case of ER.

5 Concluding practical considerations

It goes without saying that the effectiveness of all chimeras described above is only as good as their correct localization. In other words, it is important to determine that a probe specifically engineered to measure, for example, $[Ca^{2+}]$ in the

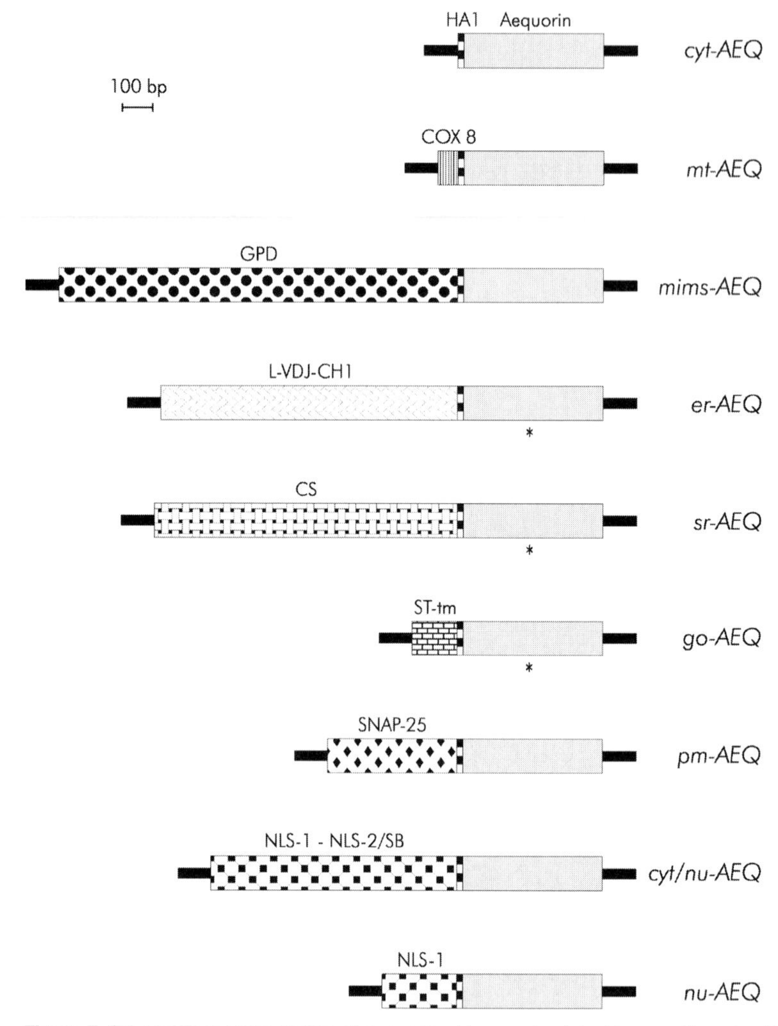

Figure 7 Schematic representation of aequorin chimeras targeted to specific subcellular locations. The chimeras represented localize to the cytoplasm (*cyt-AEQ*), the mitochondrial matrix (*mt-AEQ*), the mitochondrial intermembrane space (*mims-AEQ*), the endo- and sarcoplasmic reticulum (*er-AEQ* and *sr-AEQ*, respectively), the Golgi apparatus (*go-AEQ*), the subplasmalemma region (*pm-AEQ*), the cytosol or nucleus (depending on presence of glucocorticoids), and nucleus only (*cyt/nu-AEQ* and *nu-AEQ*, respectively). An asterisk within the aequorin portion designates the D119A mutation. Abbreviations are as follows: HA1, haemagglutinin epitope (for immunolocalization); COX 8, N-terminal fragment of subunit 8 of cytochrome *c* oxidase; GPD, glycerol phosphate dehydrogenase; L-VDJ-CH1, domains of an Igγ2b heavy chain; CS, calsequestrin; ST-tm, sialyltransferase transmembrane domain; SNAP-25, synaptosomal-associated protein; NLS, nuclear localization signal; SB, steroid binding domain.

Golgi apparatus, is indeed localized—as exclusively as possible—to this structure. Therefore, when generating a particular construct it is important to determine the exact localization of the resulting polypeptide before attempting any $[Ca^{2+}]$ measurements.

Figure 7 shows a schematic representation of all the aequorin chimeras described in this chapter. As can be seen, they all possess an HA1 domain; this is a short amino acid sequence derived from the influenza virus haemagglutinin protein. Fusion of this fragment to the C-terminal side of aequorin does not affect the photoprotein's properties, and this epitope can be readily detected by antibodies designed against the nonapeptide sequence YPYDVPDYA. Immunolocalization of the expressed aequorin chimeras was performed according to *Protocol 4*.

Protocol 4

Immunolocalization of aequorin chimeras

Equipment and reagents

- Coverslip
- Epifluorescence microscope
- Formaldehyde solution: 3.7% (v/v) formaldehyde in PBS
- PBS (see *Protocol 1*)
- Triton solution: 0.1% (v/v) Triton X-100 in PBS
- 50 mM NH_4Cl
- Gelatin solution: 1% (w/v) gelatin (type IV, from calf skin) in PBS
- Anti-HA1 mouse monoclonal antibody (12CA5, from Boehringer Mannheim)
- TRITC-labelled anti-mouse secondary antibody

Method

Cells can be processed for immunofluorescence approximately 36 h after transfection. Unless otherwise stated, all incubations are at RT.

1. Remove the culture medium and add 1 ml of formaldehyde solution.
2. Fix the cells by incubating for 20 min.
3. Wash the coverslip with the cells three times with PBS, and incubate for 10 min in 50 mM NH_4Cl.
4. Permeabilize the cell membranes by incubating in Triton solution for 5 min.
5. Incubate for 1 h in gelatin solution.
6. In a wet chamber, at 37 °C, incubate the cells with a 1:100 dilution (in gelatin solution) of the anti-HA1 mouse monoclonal antibody.
7. Wash three times with PBS.
8. In a wet chamber protected from light, incubate the cells with a 1:40 dilution (in gelatin solution) of TRITC-labelled anti-mouse secondary antibody.

Protocol 4 continued

9 Wash the cells three times with PBS.
10 Analyse the fluorescence on a standard epifluorescence microscope.

For simple verification of correct localization, a standard epifluorescence microscope is sufficient. At least for documentation purposes, however, more sophisticated systems are recommended. We generally obtain digital images with Universal Imaging's Metamorph software package, and a Zeiss Axiovert 100TV inverted microscope, equipped with a Princeton Instruments back-illuminated cooled CCD camera.

Throughout this chapter we have concentrated on general procedures. Specific applications may require minor adaptations. For example, we described some of this variety in *Protocol 3*. Other cases exist for which we have determined optimized variations to standard protocols. The two most important cases are described below.

When targeting aequorin to the mitochondrial intermembrane space, we noticed that overexpression of this chimera had deleterious effects on the transfected population. This problem was solved by reducing the quantity of DNA used for the transfection procedure. With the standard calcium phosphate protocol, we use only 0.5 μg of purified plasmid DNA (instead of the usual 4 μg).

In the case of cells transfected with an aequorin chimera destined to the subplasmamembrane region, reconstitution of active aequorin does not occur with maximal efficiency if the standard conditions described in *Protocol 2* are used. Instead, it is better to incubate the cells—at 37 °C, for 45 min—with KRB supplemented with 100 μM EGTA and 5 μM coelenterazine. After this, cells are transferred to the luminometer and maintained in KRB supplemented with 100 μM EGTA throughout the measurements.

In conclusion, recombinant aequorin is capable of being targeted to virtually any subregion of the cell. With the myriad transfection methods currently available, essentially all cell types are amenable to the use of such chimeras. Once specifically localized, this chemiluminescent probe is capable of providing precious information regarding Ca^{2+} concentration and dynamics. Specific subregions may present intrinsic difficulties, but our experience has been that by introducing minor alterations in the general procedures described herein, headway can be made.

References

1. Ridgway, E. B. and Ashley, C. C. (1967). *Biochem. Biophys. Res. Commun.*, **29**, 229.
2. Ridgway, E. B., Gilkey, J. C., and Jaffe, L. F. (1977). *Proc. Natl. Acad. Sci. USA*, **74**, 623.
3. Allen, D. G. and Blinks, J. R. (1978). *Nature*, **273**, 509.
4. Blinks, J. R., Mattingly, P. H., Jewell, B. R., van Leeuwen, M., Harrer, G. C., and Allen, D. G. (1978). In *Methods in enzymology*, Vol. 57, p. 292.
5. Cobbold, P. H. (1980). *Nature*, **285**, 441.
6. Tsien, R. Y., Pozzan, T., and Rink, T. J. (1982). *Nature*, **295**, 68.

7. Tsien, R. Y., Pozzan, T., and Rink, T. J. (1982). *J. Cell Biol.*, **94**, 325.
8. Grynkiewicz, G., Poenie, M., and Tsien, R. Y. (1985). *J. Biol. Chem.*, **260**, 3440.
9. Inouye, S., Noguchi, M., Sakaki, Y., Takagi, Y., Miyata, T., Iwanaga, S., *et al.* (1985). *Proc. Natl. Acad. Sci. USA*, **82**, 3154.
10. Cobbold, P. H. and Lee, J. A. C. (1991). In *Cellular calcium: a practical approach* (ed. J. G. McCormack and P. H. Cobbold), p. 55. Oxford University Press, Oxford.
11. Graham, F. L. and van der Eb, A. J. (1973). *Virology*, **52**, 456.
12. Rutter, G. A., Theler, J. M., Murgia, M., Wollheim, C. B., Pozzan, T., and Rizzuto, R. (1993). *J. Biol. Chem.*, **268**, 22385.
13. Gavel, Y. and von Heijne, G. (1990). *Protein Eng.*, **4**, 33.
14. Rizzuto, R., Nakase, H., Darras, B., Francke, U., Fabrizi, G. M., Mengel, T., *et al.* (1989). *J. Biol. Chem.*, **264**, 10595.
15. Rizzuto, R., Simpson, A. W., Brini, M., and Pozzan, T. (1992). *Nature*, **358**, 325.
16. Pinton, P., Pozzan, T., and Rizzuto, R. (1998). *EMBO J.*, **18**, 5298.
17. Munro, S. and Pelham, H. R. B. (1987). *Cell*, **48**, 899.
18. Sitia, R. and Meldolesi, J. (1992). *Mol. Biol. Cell*, **3**, 1067.
19. Montero, M., Brini, M., Marsault, R., Alvarez, J., Sitia, R., Pozzan, T., *et al.* (1995). *EMBO J.*, **14**, 5467.
20. Brini, M., De Giorgi, F., Murgia, M., Marsault, R., Massimino, M. L., Cantini, M., *et al.* (1997). *Mol. Cell. Biol.*, **8**, 129.
21. Robert, V., De Giorgi, F., Massimino, M. L., Cantini, M., and Pozzan, T. (1998). *J. Biol. Chem.*, **273**, 30372.
22. Marsault, R., Murgia, M., Pozzan, T., and Rizzuto, R. (1997). *EMBO J.*, **16**, 1575.
23. Picard, D. and Yamamoto, K. R. (1987). *EMBO J.*, **6**, 3333.
24. Brini, M., Marsault, R., Bastianutto, C., Pozzan, T., and Rizzuto, R. (1994). *Cell Calcium*, **16**, 259.
25. Kendall, J. M., Sala-Newby, G., Ghalaut, V., Dormer, R. L., and Campbell, A. K. (1992). *Biochem. Biophys. Res. Commun.*, **187**, 1091.
26. Barrero, M. J., Montero, M., and Alvarez, J. (1997). *J. Biol. Chem.*, **272**, 27694.

Part Two

Calcium measurement in different organelles

Chapter 5

Monitoring mitochondrial function in single cells

Mart H. Mojet, D. Jake Jacobson, Julie Keelan, Olga Vergun, and Michael R. Duchen

Physiology Imaging Consortium, Department of Physiology, UCL, Gower Street, London WC1E 6BT, UK.

1 Introduction

1.1 Historical perspective: why is mitochondrial function interesting in the context of a book on calcium signalling?

This book deals with techniques to study cellular $[Ca^{2+}]_i$ signalling. Why then include a chapter about monitoring mitochondrial function? It has been clear for many years that isolated mitochondria will take up and accumulate massive quantities of added Ca^{2+}. In the last 40 years or so, there has been a sustained curiosity about the physiological relevance of this pathway. In that time, the pendulum has swung from early suggestions that mitochondria might represent a major calcium buffering compartment, perhaps even representing a useful calcium store, to the view that mitochondria will only take up Ca^{2+} under pathological conditions, when $[Ca^{2+}]_i$ becomes dangerously high. The last 10 years or so have seen a gradual return to the recognition that mitochondria are intimately engaged in the business of calcium signalling after all—that they will take up calcium during physiological $[Ca^{2+}]_i$ signalling has now been demonstrated in a variety of cell types and this pathway has an impact at many levels. Most particularly, mitochondrial calcium uptake modulates mitochondrial function through the regulation of the enzymes of the tricarboxylic acid cycle or TCA cycle (for review, see ref. 1). Further, mitochondrial Ca^{2+} uptake regulates and helps to shape $[Ca^{2+}]_i$ signals in cells in a variety of different ways (2–4, see also ref. 5). Mitochondrial calcium uptake is of major importance in a number of pathological conditions. For example, it is now clear that mitochondrial Ca^{2+} uptake may play a decisive part in triggering cell death, especially when combined with an oxidative stress (e.g. see ref. 6).

Quite apart from their direct role in calcium uptake, mitochondria also make a major contribution to cellular Ca^{2+} homeostasis through the supply of ATP to fuel the ion pumps that maintain transmembrane gradients of Ca^{2+} and Na^+, essential determinants of cellular Ca^{2+} homeostasis. Mitochondrial integrity therefore plays a central role in the maintenance of normal cellular Ca^{2+} signalling and therefore in cell function.

The study of mitochondrial calcium handling has moved back from a marginal, peripheral study into the heart of calcium signalling—it is no accident that several large meetings on calcium signalling in 1999 include sessions on mitochondria and calcium signalling, aspects that were barely considered in this arena a few years ago. Much of this change in perspective results from application of the new technologies of fluorescence measurement and imaging that have permitted a direct study of mitochondrial function where mitochondria belong—in their native habitat of the cytosol. It therefore is possible to explore the interplay between $[Ca^{2+}]_i$ signalling and mitochondria directly within cells, and so this field of study becomes a natural area for inclusion in a book such as this. In the following pages we will describe how mitochondrial function may be assessed within cells or tissues, how those signals may be interpreted, and how to optimize the signals that you wish to study by using the most appropriate optical arrangements. Several substantial relevant reviews have recently been published which will give a much broader insight into the underlying biology (5, 7–9).

2 Fundamentals

Before going into a detailed description of the different techniques to monitor mitochondrial function, it is first essential to have a reasonable understanding of some of the basics that underlie mitochondrial function. These principles define the mitochondrial response to biochemical agents and to changes in the cellular environment, and so we will first briefly review the basic cellular physiology of mitochondrial metabolism (see *Figure 1*).

2.1 Mitochondrial metabolism

Figure 1 shows a cartoon of a mitochondrion to illustrate the chemiosmotic basis for oxidative phosphorylation. The supply of substrate, such as pyruvate, to the citric acid cycle (the tricarboxylic acid—TCA cycle or Krebs cycle) supplies reduced NADH and $FADH_2$ to the respiratory chain. Electrons from NADH or $FADH_2$ are transferred to oxygen (which is reduced to generate water) through a series of coupled redox reactions. In the process, protons are transferred across the inner mitochondrial membrane into the intermembrane space, thus generating a proton gradient which is expressed largely as a membrane potential some −150 to −200 mV negative to the cytosol. That potential, usually referred to as $\Delta\psi_m$, provides the driving force for proton influx through the ATP synthase, an enzyme complex that includes a component which is essentially a proton channel. The proton influx drives the ATP synthase which phosphorylates ADP; ATP is generated and transported to the cytosol. Some remarkable images of the operation of the enzyme were published by Noji *et al.* (10), who also posted some

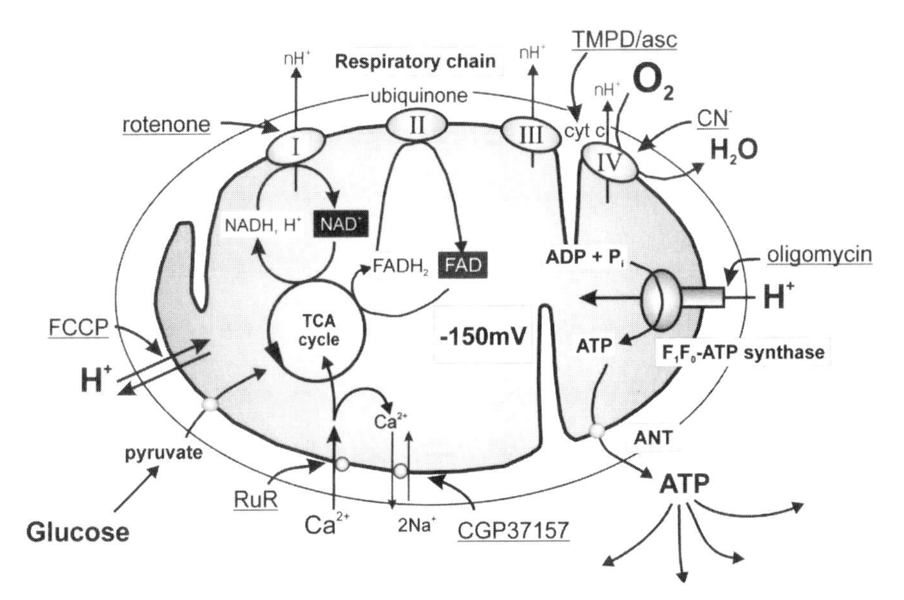

Figure 1 Cartoon of a mitochondrion to illustrate the features of the chemiosmotic basis for oxidative phosphorylation. The sites of action of some of the major biochemical reagents commonly used are underlined (see Appendix to this chapter). See text for explanation.

images on the World Wide Web (seen at http://www.bmb.leeds.ac.uk/illingworth/oxphos/atpase.htm). The mitochondrial potential also provides the driving force for Ca^{2+} uptake into mitochondria. Ca^{2+} is taken up through the uniporter when the $[Ca^{2+}]_{cyt}$ is high, moving down an electrochemical potential gradient. Re-equilibration of $[Ca^{2+}]_{mt}$ is then achieved through the $2Na^+/Ca^{2+}$ exchanger in the inner mitochondrial membrane, thought to be electroneutral (reviewed in refs 9, 11).

2.2 The relationship between $\Delta\psi_m$ and redox state

The relationship between $\Delta\psi_m$ and the redox state of NADH/NAD^+ and the flavoproteins is important to understand and may be of great experimental value. While the details of measurement will be discussed below, *Figure 2* summarizes the changes in these variables with a range of manipulations in an attempt to clarify ways in which mitochondrial potential and redox state will change with each. Thus, provision of glucose to a substrate starved pancreatic beta cell activates the TCA cycle, increasing the NADH/NAD^+ ratio (*Figure 2A*). The increased provision of reducing equivalents to the respiratory chain then leads to an increased $\Delta\psi_m$ which will in turn increase the rate of ATP production. The NADH/NAD^+ ratio is also increased by inhibition of the respiratory chain (*Figure 2B*) either by cyanide (CN^-) or by withdrawal of oxygen. The respiratory chain cannot oxidize NADH, but the rise in NADH/NAD^+ ratio is now associated with a slow depolarization of $\Delta\psi_m$—the translocation of protons by the respiratory chain stops, and protons leak either through the F_1F_0-ATP-synthase or other leak

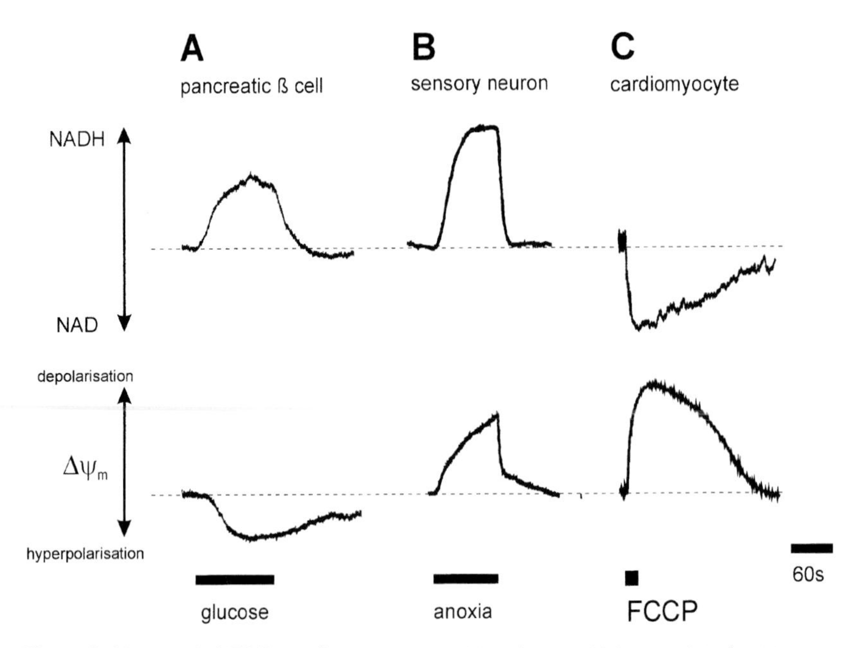

Figure 2 Changes in NADH autofluorescence and in mitochondrial potential ($\Delta\psi_m$) in response to (A) increased supply of substrate, (B) inhibition of respiration, and (C) addition of an uncoupler. Note that both (A) and (B) are associated with an increase in NADH autofluorescence, but with opposing changes in $\Delta\psi_m$, while both (B) and (C) are associated with a loss of mitochondrial potential, but with opposing changes in autofluorescence. The signals shown come from (A) a pancreatic beta cell in which the superfusate was switched from 3 to 10 mM glucose (see also ref. 53); (B) a sensory neuron in culture exposed to an anoxic perfusate; (C) a cardiomyocyte briefly exposed to 1 μM FCCP.

pathways leads to the gradual dissipation of potential (see also below). Interestingly, the apparent rate of change of $\Delta\psi_m$ under these conditions varies enormously between cell types. In contrast, collapse of $\Delta\psi_m$ using an uncoupler allows the unfettered activation of the respiratory chain—you might think of this in terms of the respiratory chain making a futile attempt to restore a potential in the face of a massive proton leak. The rate of oxygen consumption is increased, increasing the rate at which NADH is oxidized, and so now the collapse of mitochondrial potential is associated with a fall in the NADH/NAD^+ ratio (please see *Figure 2C* and Plate 1).

2.3 Consequences of mitochondrial depolarization: mitochondria as ATP consumers

Once mitochondrial respiration stops, the process generating the potential stops. What will happen next depends on the leak conductance in the mitochondrial inner membrane and the ability of the ATP synthase to reverse. The enzyme becomes a proton translocating F_1F_0-ATP-synthase, pumping protons outwards across the mitochondrial inner membrane, consuming ATP, and either maintaining $\Delta\psi_m$ or, at least, slowing the rate of dissipation. Thus, in rat cardiomyocytes, mitochondrial potential is largely sustained in the presence of CN^- until the cell becomes ATP depleted (signalled in these cells by the onset of a rigor contracture)—only then does the mitochondrial potential collapse (12). The role of the F_1F_0-ATP-synthase in maintaining $\Delta\psi_m$ is readily demonstrated by comparing the rate of change of $\Delta\psi_m$ in response to anoxia with and without the presence of oligomycin to inhibit the F_1F_0-ATP-synthase (see *Figure 1* and refs 13, 14). It seems that in some cells, particularly cell types of neuronal lineage, the mitochondrial potential dissipates more quickly than in cardiomyocytes in response to anoxia. It is never realistic to make any assumption about what is happening to $\Delta\psi_m$ with time in response to an inhibitor alone—it is necessary to measure it. However, in our experience, in response to a respiratory inhibitor and oligomycin, $\Delta\psi_m$ will collapse rapidly in all cell types. This is important—to collapse $\Delta\psi_m$ reliably can be a very useful tool in a variety of experimental situations, but one of the key problems is to separate the effects of ATP depletion from the consequences simply of the collapse of $\Delta\psi_m$. This is crucial in the interpretation of many experiments designed to examine the role of mitochondrial Ca^{2+} uptake in $[Ca^{2+}]_i$ signalling, and will be reiterated several times below.

2.4 Fluorescence measurement and imaging of mitochondrial function

We will now consider in turn the theoretical and technical issues involved in the practicalities of measuring changes in the variables that together allow us to build a clear description of mitochondrial physiology under a range of physiological and pathophysiological conditions. As major contributors to the process of mitochondrial respiration and oxidative phosphorylation, all of the following are accessible for measurement in cells:

(a) NADH redox state and flavoprotein redox state—cellular autofluorescence.
(b) Mitochondrial potential—use of potentiometric fluorescence indicators.
(c) $[Ca^{2+}]_{mt}$ (as opposed to $[Ca^{2+}]_{cyt}$)—use of fluorescent indicators or transfection with aequorin.
(d) $[ATP]_i$—indirect through measurement of $[Mg^{2+}]_i$ or by injection/transfection of luciferin/luciferase.
(e) Cytochrome redox state—cytochrome absorption spectra.

3 NADH and flavoprotein autofluorescence

3.1 Theoretical basis

The endogenous compounds that act as hydrogen carriers which ferry the protons and electrons from carbohydrates to the electron transport chain have fluorescent properties. The term autofluorescence is used to distinguish it from the fluorescence of indicators that are artificially introduced. This signal can be used as an indicator of changes in cellular metabolism, as its properties change when the carrier binds an electron. Thus the fluorescence of the pyridine nucleotide, NADH, is excited in the UV (peak excitation at about 350 nm) and emits in the blue (with a peak at about 450 nm) (see ref. 15, and *Figure 3*). The oxidized form, NAD^+, is not fluorescent. An *increase* in UV-induced blue fluorescence therefore indicates an increase in the ratio of NADH to NAD^+—a net shift in the pyridine nucleotide pool to the reduced state. It is important to emphasize that these changes in signal do not indicate net changes in the absolute size of the total pool but rather a change in the balance of reduced to oxidized forms. NADPH is also fluorescent with very similar spectral properties. NADH and NADPH are both present in both mitochondrial and cytosolic compartments, however several properties tend to mitigate in favour of the mitochondrial signal—there is more of it, the binding of NADH to membranes enhances the fluorescence while enzymatic binding tends to quench the cytosolic fraction (16).

Flavoproteins ferry electrons using a flavin or FAD molecule. Flavoprotein fluorescence is excited in the blue (peak at about 450 nm) and emission is maximal in the green, with a peak at about 550 nm (*Figure 3A*). In contrast to NADH, flavoprotein fluorescence decreases but does not completely disappear when the carrier binds electrons. A *decrease* in flavoprotein autofluorescence reflects an increase in the ratio of reduced to oxidized flavoprotein—the inverse of the response of the pyridine nucleotides (*Figure 3C* and Plate 1).

Some years ago (17) we showed that the fluorescent properties of NADH and flavoproteins are sufficiently different to allow independent measurement of the redox state of each carrier. Blowfly compound eyes were illuminated at a series of different wavelengths from UV to blue, and the emission spectra above 500 nm were recorded. The flies were then made anoxic, and the spectra were measured again. When the *difference* spectra (air *minus* anoxia) were normalized and overlaid (*Figure 3B*) it was clear that the emission spectra were isomorphic, indicating that only one molecular species is excited, by inference NADH. Likewise, the emission difference spectra of flavoproteins (*Figure 3D*) also stem from one molecular entity.

3.2 Protocols

3.2.1 Pyridine nucleotides—NADH

Since the transmission of microscope optics usually falls off rapidly below 350 nm, a good compromise between illumination intensity and excitation efficiency is obtained with a 350 ± 20 nm or 360 ± 5 nm band pass filter, and quartz optics should be used if available. Any light above 390 nm is likely to excite flavoprotein fluorescence, which will not only contaminate the NADH signal, but also result in an underestimate of the NADH signal, since flavoprotein fluorescence changes inversely with degree of reduction. The emission may be measured using a wide band pass filter with a peak at 450 nm and a bandwidth from ± 20 to 40 nm. The fluorescence tends to bleach, and excessive illumination can cause photodamage to the cell, and so, as in every case, it is best to keep the illumination intensity to a minimum consistent with reasonable signal-to-noise (please see *Figure 2* and

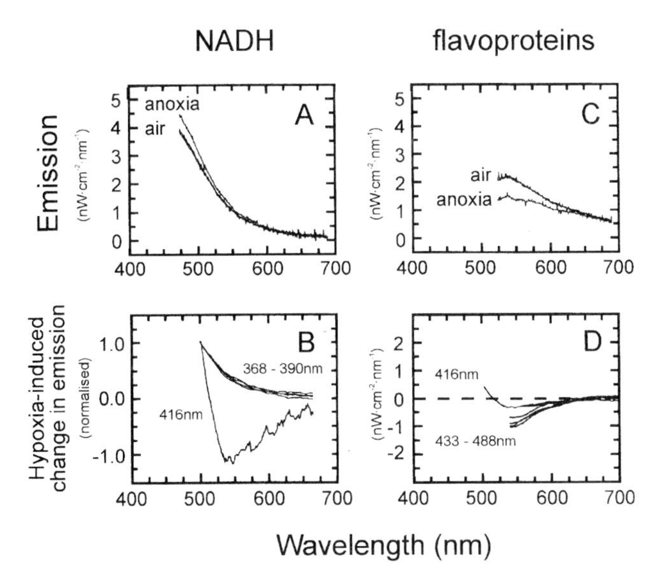

Figure 3 Spectral characteristics of autofluorescence in the blowfly compound eye. UV light (here 381 nm) excites a blue fluorescence (A) that *increases* during hypoxia. The observed fluorescence stems from one molecular species only, since a set of emission difference spectra (anoxia *minus* air) induced at excitation wavelengths between 368 and 390 nm are all isomorphic (B). This molecule is NADH, which is fluorescent in contrast to NAD^+. The fluorescence remaining in air stems partly from NADH, partly from hypoxia-insensitive fluorophores. Excitation with blue light (477 nm) induces a green fluorescence (C). This fluorescence *decreases* during hypoxia. A range of excitation wavelengths between 433 and 488 nm excites a set of isomorphic emission spectra (D), confirming the presence of only one fluorescent entity, i.e. flavoproteins. The presented autofluorescence difference spectra were obtained in a tissue particularly rich in mitochondria, the blowfly retina (17).

Plate 1). This is always maximized—in any fluorescence system—by maximizing the efficiency of the light collection pathway. Use lenses with a high numerical aperture and sensitive low noise detection systems.

3.2.2 Flavoproteins

Flavin autofluorescence is best elicited using excitation at 450 nm ± 20 nm or so and is measured between 500–600 nm (we have used a band pass filter at 550 nm ± 40 nm, or simply image with a long pass filter at > 510 nm).

For both these signals, the quality of signal tends to vary depending on the density of mitochondria within the cell. Thus, cardiomyocytes, in which the mitochondria may represent 40–50% of the cell volume, and sensory neurons, produce very large robust signals. In flat cells like astrocytes and fibroblasts which have a low overall mitochondrial volume, it may be very hard to obtain decent signal/noise to measure such signals, although good images of flavoprotein autofluorescence have been obtained with a good cooled CCD chip and long exposures.

3.3 Validation and limitations

The pyridine nucleotide and flavoprotein redox state depend on the net reactions that either oxidize or reduce them (please see *Figures 1, 2*, and Plate 1). An increase in the activity of the TCA cycle, through the Ca^{2+}-dependent up-regulation of the enzymes for example (1, 18) or through increased delivery of substrate (*Figure 2A*) will increase the balance in favour of NADH. If respiration is inhibited by mitochondrial inhibitors like CN^- or by anoxia, the respiratory chain cannot oxidize the reduced forms which will accumulate to a new steady state. Again, NADH autofluorescence will rise (*Figure 2B* and Plate 1A) and flavoprotein autofluorescence will fall (see *Figure 3* and Plate 1A). In contrast, mitochondrial respiration responds to a maximal depolarization by an uncoupler by increasing respiratory rate (see above). This leads to maximal oxidation of NADH to NAD and $FADH_2$ to FAD, decreasing the autofluorescence from NADH and increasing that from FAD (*Figure 2C* and Plate 1C and see above, Section 2).

It is difficult to estimate the full range of autofluorescence, i.e. when an electron carrier is completely oxidized or reduced. It is usually assumed that bath application of CN^- stops all electron flux completely, leaving all NADH and flavoprotein in the reduced state; this assumption is valid as long as most NADH is contained in the mitochondria, and relatively little cytosolic NADH is oxidized by lactate dehydrogenase. Application of uncouplers like FCCP will maximize O_2 consumption, and therefore electron flux. However this does not necessarily mean that the NADH is completely oxidized, as dehydrogenases will continue to supply reducing equivalents. Only when glycolysis and fatty acid oxidation are completely blocked can it be assumed that the NADH autofluorescence is minimal. Unfortunately, inhibition of glycolysis by iodoacetate or deoxyglucose are irreversible. It should be clear that an increase in NADH autofluorescence signal can be interpreted in different ways—either an increase in effective substrate supply or an inhibition of oxidation, and other experiments may be necessary to differentiate between these.

It is important to understand these changes in autofluorescence, not only for their own sake, but also because these changes may contaminate signals from other fluorescence indicators—perhaps most notably Indo-1, and will cause changes in raw fluorescence signals in response to metabolic manipulations which have nothing to do with changes in $[Ca^{2+}]_i$.

4 Fluorimetric measurement of mitochondrial potential

4.1 Theoretical basis

The extrusion of protons across the inner mitochondrial membrane from matrix to intermembrane space by the electron transport chain generates a proton gradient across the inner mitochondrial membrane mostly expressed as a potential of the order of -150 to -200 mV or so. This potential ($\Delta\psi_m$) lies at the

Table 1 Fluorescent dyes commonly used in the study of mitochondrial physiology

	Stock solvent[a]	Loading concentration[b]	Surfactant required?[c]	Loading time[d] (min)	Peak excitation[e] (nm)	Peak emission[e] (nm)
Fura-2 AM	DMSO	5 μM	Yes	30	350/355	540
Fluo-3 AM	DMSO	4.4 μM	Yes	30	490	530
Indo-1 AM	DMSO	4.95 μM	Yes	30	*348*	*485/400*
Rhod-2 AM	DMSO	4.4 μM	Yes	30	555	600
TMRE/TMRM	DMSO	5 nM/1.5 μM	No	15	555	590
Rh123	H_2O	200 nM/2.6 μM	No	15	*510*	*530*
JC-1	DMSO	10 μM	No	15	*515*	*530 / 590*
DASPMI	Dimethyl-formamide	20 μM	No	5	550	*461*
$DiOC_6$	DMSO	40 nM	No	5	*484*	*501*

[a] The commercial product is usually diluted to a stock solution (concentration around 1 mM) which is stored at −20 °C. The final dilution for cell loading is done in physiological saline.

[b] Potentiometric probes (TMRE, TMRM, and Rh123) may be loaded at a low concentration (redistribution mode) or a high concentration (quench mode) respectively.

[c] The solubility of lipophilic AM esters in water is low and the dispersant Pluronic® F-127 (stock solution: 16 mM in DMSO, stored at room temperature; final concentration 40 μM) may be required. Cationic fluorophores are hydrophilic and do not require dispersant.

[d] Dye loading was done at room temperature (see text, Section 5.2). Dye loading is normally followed by washing of the cells with physiological saline to remove extracellular dye and reduce background fluorescence. However, in the case of the potentiometric dyes used in 'redistribution mode', the cells are not washed and the 'loading time' reflects the time taken for the dye to equilibrate from solution into mitochondria.

[e] These peak excitation and emission wavelengths are based on measurements performed in our laboratory using dye-loaded astrocytes, a grating-based monochromator, and a grating-based spectrograph with CCD detector. The wavelengths may vary slightly from those quoted in the literature (which are often obtained from *in vitro* measurements where the dyes are dissolved in methanol) and will vary according to the transmission optics of the microscope used and the sensitivity of individual detectors. The wavelengths in *italics* were obtained from the Molecular Probes website (http://www.probes.com/).

heart of mitochondrial oxidative phosphorylation—it is harnessed by the mitochondria to drive uptake of otherwise impermeant ions across the inner mitochondrial membrane, notably calcium and ADP and the downhill movement of protons drives the F_1F_0-ATP-synthase (19).

In isolated mitochondria, for many years the potential has been measured by following the distribution of lipophilic cations such as tetraphenyl phosphonium (TPP^+) using an electrode and a semi-permeable membrane. The same principle has been employed, using fluorescent lipophilic cations. The existence of a single, delocalized charge on the fluorescent compounds rhodamine-123 (Rh123), tetramethyl-rhodamine ethyl and methyl esters (TMRE and TMRM, respectively), 5,5′,6,6′-tetrachloro-1,1′,3,3′-tetraethylbenzamidazolocarbocyanine (JC-1) and $DiOC_6(3)$ and DASPMI (see *Table 1*) permits these dyes to cross the cell membranes easily, where they partition into compartments in response to their electrochemical potential gradients. Most of these indicators partition between the perfusate and cytosol and then between the cytosolic and mitochondrial compartments in response to a series of potential differences.

4.2 Approaches to the use of indicators and technical requirements

4.2.1 Use of indicators in fluorescence 'quench and dequench' mode

When we and others started this kind of work, in the late 1980s, imaging technology was not generally accessible and we used photomultiplier tubes to measure the averaged signal across a cell or small cluster of cells. Under these conditions, if a dye moves from the mitochondria to the cytosol in response to a mitochondrial depolarization, it is quite possible to measure no change in signal at all (please see Plate 3Biii). This was the case with the dye DASPMI, which we tried using in 1989 and abandoned, as we could see no change in signal in response to an uncoupler. In fact, DASPMI although little used, is probably a useful dye, but only when imaging at high spatial resolution. Several groups therefore adopted a strategy which may seem confusing: dyes were loaded into cells at relatively high concentrations (~ 10–20 μM) for short periods—usually 10–15 minutes—followed by washing. At higher concentrations, many dyes undergo a phenomenon called autoquenching—energy is transferred by collisions between monomeric dye molecules. The concentration of dye may also promote the formation of aggregates of dye molecules which may be non-fluorescent. This tends to occur following the potential-dependent concentration of indicator into the mitochondria. Mitochondrial depolarization promotes the redistribution of the dye into the cytosol where its dilution relieves the quench and the net fluorescence signal increases (please see Plate 2). Thus, in this mode, mitochondrial depolarization is associated with an increase in fluorescence. TMRM, TMRE, and Rh123 have all been used in this mode by several groups.

4.2.2 Use of dyes as redistribution probes

An alternative approach, theoretically more satisfactory, is to bathe cells in a very low concentration of dye—in the range of 2–30 nM (e.g. see ref. 20). Concentrations up to 300 nM have been used, but the higher the concentration the greater the chance of quench and non-specific binding, varying with the indicator: see *Table 1*. Under these conditions, in the continuous presence of the indicator, it will equilibrate between saline, cytosol, and mitochondria. Mitochondrial depolarization causes the redistribution of dye from mitochondria to cytosol and very little net change in signal over the whole cell (please see Plate 3), whilst depolarization of the plasma membrane may cause loss of dye from the cytosol across the plasma membrane. To use this technique well, high resolution microscopy, ideally confocal microscopy, is required, with digitization to 12- or even 16-bits if it is to prove possible to resolve differences in cytosolic and mitochondrial signals which may reach a difference of several hundred-fold. It is interesting to note in passing how mitochondria may look quite different in different cell types. For example, they tend to be 'bullet'-shaped and squat in neonatal cardiomyocytes in culture (Plate 3A), much more elongated and rod-like in fibroblasts and astrocytes (Plates 2A and 3B), and rather smaller and more difficult to identify discretely in neuronal somata (please see Plate 5A).

4.2.3 JC-1: a ratiometric indicator?

This dye is marketed as a 'ratiometric indicator' of mitochondrial potential. This has an immediate appeal as it holds promise of standardization and quantification of data. The principle of the indicator is that at high concentrations, the dye forms complexes (J-complexes) which show a striking red shift in the fluorescence emission spectrum—with excitation at ~ 490 nm, the peak emission of the monomer is at ~ 539 nm while the J-complex emission peaks at 597 nm, in the red. The concentration of dye into mitochondria therefore should promote aggregation, and so the greater the potential, the more red signal we should see. Dual emission fluorimetry, with continuous measurements at ~ 539 and ~ 597 nm should provide a ratio measurement of $\Delta\psi_m$. Indeed, we have found that an uncoupler tends to decrease the red signal and increases the green, although these typically change at different rates. However, experience in different laboratories seems to show a substantial variability. Di Lisa *et al.* (21) reported that J-aggregates giving rise to red fluorescence appear in the aqueous phase, while the green fluorescence of the monomer requires a lipid phase and therefore may reflect the accumulation of dye in membranes. They also found that the red fluorescence signal appeared more sensitive to relatively small changes in signal than the green signal, which only changed significantly with larger excursions of $\Delta\psi_m$.

Certainly, the red form appears sensitive to factors other than $\Delta\psi_m$—a recent paper by Chinopoulos *et al.* (22) showed that H_2O_2 had profound effects on the red fluorescence independent of any change in $\Delta\psi_m$. Further, the equilibrium between monomers, dimers, and polymers is not solely due to membrane potential, as JC-1 is pH-sensitive and its absorption spectrum may be affected by the osmolarity of its environment (23).

Confocal imaging of JC-1 generates some quite strikingly beautiful images, but interpretation is very difficult. In some preparations, some mitochondria appear red, some green—does this mean that there are populations of mitochondria with different potentials? We, and others, have even seen single mitochondria that are mostly green but have some red sections to them (see also ref. 24). But how can we best decide how to interpret these signals? Careful confocal imaging of the same cell types with very low concentrations of TMRE in our hands have so far failed to reveal the same differences in population seen with JC-1 and so we remain very cautious about the interpretation of JC-1 signals. There are some instances where this choice of dye seems to be the best option (i.e. in comparing populations of cells subjected to different regimes), but have generally chosen wherever possible to use an alternative dye as we lack confidence in changes in JC-1 signal as reliable indicators of changing $\Delta\psi_m$.

4.3 Calibration?

As an ion will partition across a polarized membrane according to the Nernst equation, one might expect a lipophilic dye bearing a charge to behave likewise, as long as the concentration is very low, with no quench and no non-specific binding.

Hence, from:

$$V = -(RT/ZF) \ln (c_i/c_o)$$

the term c_i/c_o (ratio of intracellular concentration over extracellular concentration of the ion under study) may be replaced with one of F_m/F_o, where F_m is the fluorescence over the mitochondrion and F_o is that over the mitochondrion-free cytosol adjacent to the organelle (the fluorescence ratio must be collected confocally to exclude fluorescence from out-of-focus regions of the cell) again with the proviso that the fluorescence signal must be proportional to dye concentration.

In practice, however, most cationic dyes show significant binding to lipid membranes, and membrane binding can enhance the fluorescence of the probe. Thus, the distribution of the fluorescence signal may not be strictly Nernstian and it is therefore very difficult to provide reliable calibrations for the signals from these dyes to quantify $\Delta\psi_m$, although some have succeeded in putting what look like sensible numbers next to their images. However, useful qualitative information on mitochondrial polarity is reflected by its distribution and thus valuable —and otherwise inaccessible—data may be obtained by digital imaging of dye-loaded cells. These issues are dealt with in some depth by Fink *et al.*, 1998 (25).

4.4 Validation

Imaging of cells stained with most of these fluorophores reveals obvious loading into mitochondria. Using JC-1 we have seen red fluorescent objects which are not mitochondria (including the nuclei of dead cells!). There is a tendency for reviewers to ask for validation using 'mitotracker' dyes, but there is a circularity of argument here, as the mitotracker indicators typically load preferentially into mitochondria in exactly the same way as the potentiometric indicators, and so of course the indicators co-localize.

One must be careful to address several concerns, for each dye and each cell type that you use:

(a) Does the signal change reflect only changes in mitochondrial potential or is there a substantial change with depolarization of the plasma membrane?

(b) How quickly and reliably does the dye follow changes in $\Delta\psi_m$ in response to a range of manipulations that should alter $\Delta\psi_m$ predictably?

(c) Is there any evidence of toxicity of the dye in terms of cell function, mitochondrial function, etc.?

4.4.1 Plasma membrane potential

Perhaps the most rigorous way to examine the effect of plasma membrane depolarization is to record changes in signal while voltage-clamping the cell membrane potential. Using this approach, we showed that a voltage step from −70 mV to +60 mV (the potential at which the Ca^{2+} current appears to reverse) caused no significant change in rhodamine-123 signal, while a step to 0 mV caused a small but significant (~ 20%) increase in signal (18). This latter response was shown to be due to the mitochondrial response to a rise in $[Ca^{2+}]_i$ and mitochondrial calcium uptake (which generates an inward current across the mitochondrial inner membrane and a transient depolarization). Whole cell patch clamping as an approach to studying metabolic processes is always problematic, as $[Ca^{2+}]_i$ is usually buffered and ATP need be given in the pipette solutions to avoid washout of currents. An alternative is to use the amphotericin perforated patch technique (e.g. see ref. 26).

Depolarization is easily achieved using high potassium (50 mM will give a depolarization close to 0 mV), and this can be done with and without Ca^{2+} to establish the component of the response due simply to membrane depolarization. This is discussed in some detail by Rottenberg and Wu (26), but in our hands, using Rh123 or TMRE in dequench mode or using TMRE at very low concentrations, we see no significant change in signal simply with plasma membrane depolarization, even if one might expect it on theoretical grounds. $DiOC_6(3)$, often used as a probe of mitochondrial potential has also been used as a probe of plasma membrane potential, and great care must be taken in interpreting signals with this dye (see ref. 27).

4.4.2 Manipulation of mitochondrial membrane potential

The typical published study employs (at most) a mitochondrial uncoupler to demonstrate the change of signal with mitochondrial depolarization. While this is not unreasonable as a starting point, it hardly suffices to validate the use of the dye in full. One approach is to use a preparation of isolated mitochondria and to look in parallel at the changes in fluorescence signal and at a TPP^+ electrode trace. Accumulation of TMRE or Rh123 in respiring mitochondria will cause a quench of the dyes that faithfully follows the change in TPP^+ electrical signal (Duchen, Leyssens, and Crompton, unpublished). We would also like to suggest the following scheme, illustrated in *Figure 4*.

(a) Apply rotenone (1 μM), which inhibits respiration at complex I. Depending on cell type, this causes a slow depolarization or none at all. If the depolarization is small, it may be interesting to test whether the F_1F_0-ATP-synthase contributes to maintaining $\Delta\psi_m$, acting as an ATPase (in 'reverse mode') and using glycolytic ATP.
   (i) Addition of oligomycin (2.5 μg/ml) after rotenone is washed out may cause a small hyperpolarization of $\Delta\psi_m$ or no change at all (the hyperpolarization suggests some resting proton flux through the F_1F_0-ATPase which is causing a tonic depolarization).
   (ii) Application of rotenone in the continued presence of oligomycin should now cause a faster and larger depolarization, as the F_1F_0-ATPase cannot operate in 'reverse mode' to maintain the potential.

(b) In the presence of rotenone, application of TMPD (~ 10 μg/ml) and ascorbate (~ 1 mM: see Appendix to this chapter) should restore the potential,

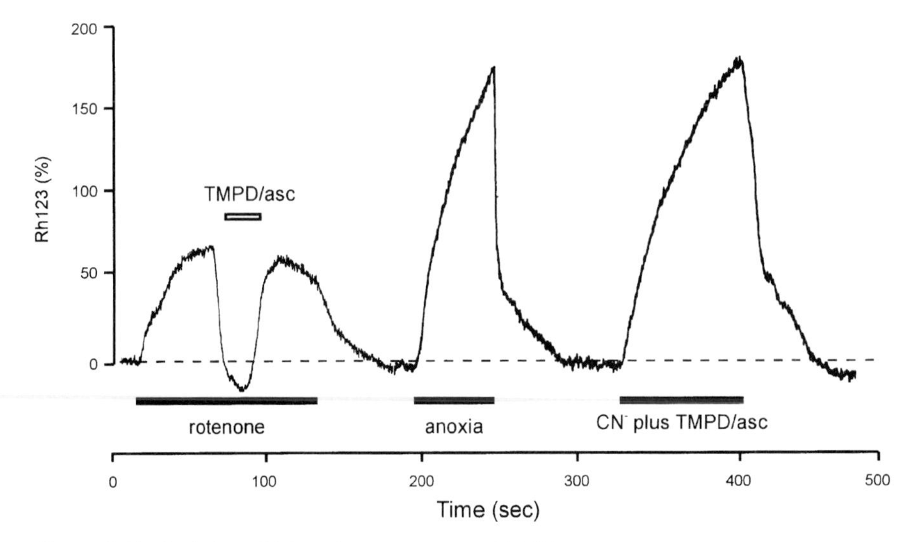

Figure 4 Manipulation of $\Delta\psi_m$ using a sequence of inhibitors and electron donors. This record comes from a mouse sensory neuron in culture loaded with Rh123. Rotenone was applied, as indicated, and TMPD with ascorbate (asc) added at the peak of the response. The TMPD donates electrons directly to complex IV, bypassing the block by rotenone at complex I. Anoxia and CN^- have similar effects on $\Delta\psi_m$, as both block at complex IV. The response to CN^- was not affected at all by the presence of TMPD/asc, as it blocks beyond the site of donation by the TMPD. TMPD/asc had almost no effect at all on the resting signal in these cells, but can stimulate respiration and increase the potential. See text for further explanation.

bypassing the inhibition by rotenone and delivering electrons directly to cytochrome *c*.

(c) A further addition of cyanide (1 mM) will depolarize the potential again, despite the presence of TMPD/ascorbate, as CN^- blocks downstream of the TMPD site (*Figure 4*).

When we tried this sequence of manipulations using JC-1 we routinely saw a signal that would be interpreted as mitochondrial depolarization with oligomycin. Either this means that the mitochondria were damaged and were maintaining a potential with the F_1F_0-ATP-synthase or that the JC-1 responded directly to the drug in some way. It is worrying. The procedure generates predictable and reliable responses using Rh123 and TMRE or TMRM in dequench mode. Using TMRE at a very low concentration (10 nM), and confocal microscopy, one can detect the repeated movement of dye between cytosol and mitochondria with these manipulations.

4.4.3 Some anomalies?

Rh123 has 'worked' well as an indicator of $\Delta\psi_m$ in all cell types that we have tried, except in cardiomyocytes. In these cells, the distribution of dye is clearly mitochondrial but uncouplers cause no change in signal. TMRE seems to behave in cardiomyocytes exactly as it does in all other cells. We have no idea what is the basis for these differences. JC-1 may also be satisfactory in cardiomyocytes, but seems slow compared to TMRE, and may be unreliable, as discussed above.

4.5 Toxicity

4.5.1 Direct toxic effects

Most indicators have direct toxic effects on mitochondrial function that are usually ignored. Thus, carbocyanine dyes such as $DiOC_6(3)$ and JC-1 have long been known to inhibit complex I (NADH dehydrogenase) and at concentrations of 40–100 nM, concentrations often used in flow cytometric studies, $DiOC_6(3)$ inhibits mitochondrial respiration by $\sim 90\%$, equivalent to rotenone! Rh123 inhibits the F_1F_0-ATPase at high concentrations (28).

4.5.2 Photodynamic toxicity

Photobleaching of fluorescent dyes, whereby light-induced oxidation of the fluorophore results in a gradual loss of signal, may occur if intensity of illumination by the excitation light is too great. This is particularly relevant if dyes are loaded at low concentrations when low fluorescence signals may prompt an increase in excitation intensity. If this is unavoidable, limiting the duration of exposure by closing a shutter intermittently may reduce bleaching.

Light-induced oxidative damage can be equally problematic with mitochondrial probes loaded at high concentrations. Illumination of fluorophores may result in the production of reactive oxygen species, and the consequent oxidative stress can have wide-ranging effects, including induction of local calcium release, inhibition of the electron transport chain, and induction of mitochondrial permeability transition. As with photobleaching, limiting the intensity of the excitation light by attenuating the excitation output or reducing the period of illumination will reduce the oxidative stress.

4.5.3 Other problems

Many of the lipophilic cations used for these measurements are substrates for the multidrug resistance transporter (MDR)—indeed, some have been used to try to quantify MDR activity, and so some cells may actively export the dyes which cannot equilibrate according to the principles set out above. Also, it is important to note that cyclosporin A (CsA) is an inhibitor of the MDR. Effects of CsA on fluorescence from these dyes must therefore be treated with great caution, and other inhibitors of the MDR such as verapamil, should also be tested.

4.6 An interesting question?

A common observation now in measurements of $\Delta\psi_m$ is that some manipulation will produce a change in $\Delta\psi_m$. That change might be moderate as judged by the change in signal in response to an uncoupler, say a 40% increase in the signal from Rh123. But what does that mean? Does it signify a mean depolarization of all mitochondria by about 40%? Does it mean a complete collapse of some mitochondria whilst others are fully intact? Similarly, we may see a gradual depolarization of mitochondria in a cell over time. What is happening with time? Are all mitochondria depolarizing slowly? Or is each individual mitochondrion undergoing a complete collapse of potential but staggered over time, so that the net response from a population looks 'smeared' in time? Certainly, this latter phenomenon may occur—this was beautifully demonstrated by Hüser *et al.* (29) in a preparation of isolated mitochondria plated as a lawn onto a coverslip and then imaged using confocal microscopy. Individual mitochondria underwent step-like transitions to complete loss of potential, but the mean response of the population showed a gradual slow depolarization. If these questions are to be answered in cell systems, then high resolution confocal microscopy will surely be an absolute requirement.

5 Measuring mitochondrial calcium ($[Ca^{2+}]_{mt}$)

5.1 Introduction and theoretical basis

The measurement of changes in mitochondrial calcium ($[Ca^{2+}]_{mt}$) is of wide interest in physiology and pathophysiology. The use of membrane-permeant, esterified fluorescent calcium indicators to measure $[Ca^{2+}]_{mt}$ in preparations of isolated mitochondria is relatively straightforward. However, the selective measurement of $[Ca^{2+}]_{mt}$ in the mitochondria of intact living cells is more difficult, and studies on the role and regulation of mitochondrial Ca^{2+} transport in cells were hampered for many years by the lack of available techniques to make such measurements. There are in principle two approaches to assessing the role played by mitochondria in the calcium homeostasis of the cell; one can prevent mitochondrial Ca^{2+} uptake and examine the consequence for Ca^{2+} signalling; alternatively one can measure mitochondrial calcium uptake directly. The two approaches of course give different information and are nicely complementary.

Direct measurement of $[Ca^{2+}]_{mt}$ *in situ* is now also possible using a technique pioneered by Rizzuto and his colleagues in the early 1990s, whereby cells are transfected with chimeric cDNA encoding the Ca^{2+}-sensitive photoprotein aequorin. A mitochondrial presequence from the mitochondrial enzyme cytochrome oxidase is also encoded in order to specifically target the fusion protein to mitochondria. This is an elegant technique ensuring specific measurement of Ca^{2+} changes in the discrete mitochondrial compartment, although the fluorescence of aequorin is too low to be useful for $[Ca^{2+}]_{mt}$ measurement at the single cell level using current technology, and data are gathered from cells in suspension. This technique demands mention in a section such as this, but is discussed in detail in Chapter 4.

In this section, we will consider the use of fluorescent dyes. The contribution of $[Ca^{2+}]_{mt}$ to shaping cellular Ca^{2+} dynamics can be determined by estimating the changes in cytosolic Ca^{2+} ($[Ca^{2+}]_{cyt}$) in the presence and absence of blockade of mitochondrial calcium uptake, using fluorescent Ca^{2+} indicators confined to the cytosol to measure $[Ca^{2+}]_{cyt}$. More direct estimates of changes in $[Ca^{2+}]_{mt}$ can be made:

(a) By manipulating the loading protocols of esterified fluorescent dyes such that preferential accumulation into mitochondria is achieved.

(b) By loading cells with high levels of a fluorescent calcium indicator followed by quenching of the fluorescent signal from the cytosolic dye by Mn^{2+} ions, leaving only the mitochondrially located fluorescence (30).

(c) By removal of the cytosolic component using whole cell patch clamping to washout the cytosol dye (31, 32).

(d) Using dyes which tend to localize to the mitochondria—e.g. rhod-2 AM.

5.2 Protocols

5.2.1 Imaging $[Ca^{2+}]_{mt}$ in single mammalian cells by selective loading of fluorescent indicators

When single cells are loaded with standard acetoxymethyl ester (AM) derivatives of fluorescent indicators, the dyes inevitably localize into both organelle and cytosolic compartments, although the relative degree of localization differs between cell types and between dyes and is dependent on dye concentration and loading temperature. A combination of certain cell types and fluorescent indicators can provide mitochondrial loading with little effort. In many cells, standard loading techniques for dyes such as fura-2 may result in as much as 50% of dye within organelles. This can be exploited by exaggerating those aspects of the loading procedures that increase compartmentalization and encouraging dye loss from the cytosol. For example Csordás *et al.* (33) found that a mast cell line (RBL-2H3 cells) loaded with 5 μM fura-2FF/AM at room temperature for 60 minutes yielded a predominantly mitochondrial signal (33). Similarly, using a high concentration of Indo-1 (10 μM Indo-1 AM for 30 minutes at room temperature) promoted the accumulation of the dye into organelles, including mitochondria, in guinea-pig ventricular myocytes (34).

In other instances, parameters of dye concentration and temperature need to be manipulated to a much greater degree in an attempt to get more selective loading into mitochondria. For example, Griffiths *et al.* (35) found that isolated rat ventricular myocytes can be loaded with Indo-1 AM under conditions in which about half the dye is located within mitochondria (5 μM, 15 minutes at 30 °C). If cells were then 'heat-treated' (1.5 hours incubation at 37 °C), only the mitochondrial fluorescence remained. These protocols exploit the temperature sensitivity of plasmalemmal anion transporters to remove the cytosolic fraction of the indicator, and clearly depend on the expression of the transporters by the cell type being studied.

The current dogma seems to be that loading of esterified indicators (typically 5 μM) into cells for short periods (15–30 minutes) at higher temperatures (37 °C) preferentially yields a cytosolic localization because the activation of cytosolic esterases cleaves ester moieties from the dye prior to its passage into organelles, and thus traps the free acid form in the cytosol. Loading at lower temperatures (often room temperature, although some researchers recommend 4°C) for longer periods allows a greater degree of dye loading into organelles. The extent to which protocols involving exposure to higher temperatures for the selective removal of cytosolic dye are applicable to the wide array of fluorescent Ca^{2+} indicators and cell types remains to be determined.

5.2.2 Removal of cytosolic fluorescence with the Mn^{2+} quench method

The judicious use of fluorescent Ca^{2+} indicators and various loading conditions have frequently been employed to monitor $[Ca^{2+}]_{mt}$ using traditional imaging techniques, such as CCD cameras. However the use of such systems, equipped with conventional microscopes, results in poor image resolution which is often insufficient to discern organelles such as mitochondria, and $[Ca^{2+}]_{mt}$ measurements are probably contaminated by cytosolic fluorescence. In an attempt to enable more precise measurement of $[Ca^{2+}]_{mt}$ by traditional imaging techniques Miyata *et al.* (30) loaded single cardiomyocytes with a high concentration (10 μM or greater) of Indo-1, and then exposed cells to Mn^{2+}. If Mn^{2+} enters the cell, it binds with high affinity to the Indo-1 and quenches the fluorescence. This has been exploited to quench cytosolic fluorescence, leaving the measured intensity as a reflection of mitochondrial Ca^{2+}.

The Mn^{2+} quench method cannot be applied to all cells however. Miyata *et al.* (30) found that cytosolic fluorescence is quenched rapidly upon superfusion with Mn^{2+} in rat myocytes, and the fluorescence signal then remains stable for some time. However, Mn^{2+} is also sequestered by mitochondria via the mitochondrial Ca^{2+} uniporter (36), and in guinea-pig and rabbit myocytes a rapid quenching of Indo-1 in mitochondria renders the Mn^{2+} quench method inappropriate for the measurement of $[Ca^{2+}]_{mt}$. Furthermore, intramitochondrial Mn^{2+} inhibits the efflux of Ca^{2+} from mitochondria. Mn^{2+} also depletes cellular energy supplies by interfering with oxidative phosphorylation at the level of the F_1F_0-ATPase and complex I (36). The Mn^{2+} quench method is now little used and perhaps should only be applied with caution and relevant controls to ensure that cell physiology is not otherwise unduly affected by this method.

5.2.3 Rhod-2

The AM ester of rhod-2 is currently the only cell permeant Ca^{2+} indicator that carries a delocalized positive charge, thus promoting preferential uptake into the negatively polarized mitochondria. Upon hydrolysis of ester moieties the rhod-2 free acid remains trapped inside the mitochondria where it reports increased $[Ca^{2+}]_{mt}$ as an increase in fluorescence intensity. In our hands, in almost all cell types, the dye effectively partitions between cytosol and mitochondria so that a significant proportion of dye is left in the cytosolic compartment. The protocol necessary for loading cells with rhod-2 is shown below.

Protocol 1

Loading cells with rhod-2

Reagents

- Rhod-2 AM ester (supplied freeze-dried in aliquots from Molecular Probes)
- Pluronic, dissolved in DMSO to a concentration of 2%
- Dry DMSO
- A standard physiological Hepes-buffered saline

Method

1. Reconstitute the rhod-2 AM with the addition of 50 μl of dry DMSO and agitate to mix well. Once reconstituted, the stock solution should be aliquoted out to small volumes and kept frozen until needed.
2. To load the cells, we use 1 ml of saline to which are added 5 μl of the rhod-2 stock together with 2.5 μl of the Pluronic stock, giving final concentrations of 4.5 μM and 0.005% respectively.
3. Take the coverslip of cultured cells, simply remove the culture medium, wash gently with saline, and then replace the solution over the cells with the loading solution. Cells are left on the bench at room temperature to load for 20–30 min—the time required for adequate loading may vary with cell type.
4. Remove the loading solution and wash gently two or three times with saline before imaging.
5. The mitochondrial localization of this dye can be further enhanced in some cell types by incubating the cells at 37 °C (after washing) for several hours or overnight (37). To follow such a protocol the culture medium must be retained, after removal in step 3, and kept in the incubator to avoid temperature and pH changes until it is put back onto the cells for a prolonged incubation period. During this incubation period plasma membrane transporters may eliminate the residual cytosolic dye without affecting the mitochondrial loading. Similar prolonged incubations apparently enhance the mitochondrial localization of rhod-2 in myocytes (38), hepatocytes (39), and oligodendrocytes (40).[a]

[a] Care should be taken not to apply such protocols blindly to all cell types however, because in our hands hippocampal neurons became significantly less responsive to glutamate after an overnight incubation to try and improve rhod-2 localization.

Dihydro-rhod-2 AM is a non-fluorescent derivative, readily made in the laboratory from rhod-2 AM (instructions provided by Molecular Probes). This is taken up into mitochondria, in the same way as rhod-2, and the cells are then washed and left in the incubator for some hours or overnight. Oxidation of the dye within the mitochondria produces the fluorescent rhod-2. We have tried this protocol and see beautiful and clearly selective loading of dye into mitochondria. However, we found these signals completely unresponsive to manipulations that raise $[Ca^{2+}]_i$ massively, and have ceased using this approach. The lack of literature using the dye this way suggests that others have also found it difficult to use, and it does raise questions about the performance of the dye.

Conventional imaging techniques (such as a CCD camera) can be readily applied to some cell types as long as they are fairly flat—for example astrocytes (4), in which mitochondria can be easily resolved. Other cell types, including 'fatter' neurons require the use of confocal microscopy to resolve the rhod-2 fluorescence in different compartments unequivocally, as the mitochondrial signal may be completely swamped by the larger volume of cytosolic signal.

5.2.4 Using confocal microscopy to measure $[Ca^{2+}]_{mt}$

Confocal microscopy will permit improved resolution of the mitochondrial rhod-2, and reduce contamination by cytosolic fluorescence. Of course the confocal microscope has its limitations. For example, the confocal optical slice can be thicker than the diameter of the mitochondria, and it becomes difficult to be certain that the fluorescence signal from the mitochondria is completely free from the signal contributed from overlying cytosol, although it will be significantly less than with conventional imaging techniques. Peng *et al.* (41) have commented that measuring the fluorescence from a region of interest which includes large areas of cytoplasm outside the mitochondria will 'dilute' the changes in $[Ca^{2+}]_{mt}$, for example in response to NMDA (41), but if the cytosol also contains a significant amount of dye then the measured signal will not be diluted but rather contaminated.

It is often a challenge to follow the fluorescence of individual mitochondria using confocal microscopy because mitochondria tend to move about in time, at least in some cells types. On this note, we have encountered quite substantial problems concerning the movement of mitochondria in and out of the confocal plane whilst studying changes in $[Ca^{2+}]_{mt}$ in response to glutamate exposure in hippocampal neurons, and these problems are exacerbated by the substantial cell swelling which occurs under such conditions. In an attempt to address such problems we now co-load neurons with rhod-2 and the mitochondrially located dye mito tracker green (MTG), and ratio the fluorescent signal from these dyes. If both signals disappear, then the mitochondrion has moved. If the rhod-2 signal falls while the MTG signal remains, then the mitochondrion has lost Ca^{2+} (or dye, or both). This approach has also been adopted in confocal imaging of hepatocytes (37).

5.3 Validation and limitations

An important point to highlight when discussing the validations and limitations involved in measuring $[Ca^{2+}]_{mt}$ is that one cannot assume that dyes marketed as 'cytosolic' or 'mitochondrial' actually localize to these compartments when loaded as AM esters, and that adequate controls need to be conducted to be sure that one knows the true origin of the fluorescence signal in a particular preparation. This is particularly important when using conventional imaging techniques where mitochondria may not be adequately resolved.

Fluorescence can be shown to originate primarily from mitochondria if the fluorescence signals can be altered by the following methods:-

(a) Inhibition of mitochondrial Ca^{2+} uptake and efflux by ruthenium red and clonazepam, respectively (see Appendix to this chapter). Ruthenium red is a large cationic molecule that should by rights be impermeant. In fact, it has been reported as entering some cells (e.g. cardiomyocytes), but it generally requires specific methods of introduction into the cell, such as microinjection or lipofusion techniques. When microinjected, its effects on the plasma membrane are avoided. A newly marketed 'cell-permeant' derivative—RU360 (42) is now available, although in our hands this compound altered physiological intracellular Ca^{2+} signalling in astrocytes (Jacobson and Duchen, unpublished observations).

(b) Inhibition of mitochondrial Ca^{2+} uptake by the application of a mitochondrial uncoupler, such as FCCP, to abolish $\Delta\psi_m$. Caution is needed with this approach however, to ensure that any observed changes are not due to a secondary effect on cellular energy levels and the consequential inhibition of Ca^{2+} efflux via the plasma Ca^{2+}-ATPase. Inclusion of low concentrations oligomycin (e.g. 2.5 μg/ml) to inhibit any potential ATP hydrolysis via the F_1F_0-ATPase can help address this issue. Collapse of $\Delta\psi_m$ may similarly be achieved using a respiratory inhibitor together with oligomycin, and this might help ensure that effects of uncoupler are not due to collapsing proton gradients in other compartments or to pH shifts.

Even with the judicious use of mitochondrially located fluorescent dyes, there can still be limitations in the application to some models. For example, just as high affinity cytosolic dyes such as fura-2 and fluo-3 may underestimate large $[Ca^{2+}]_{cyt}$ changes, rhod-2, with a K_D for Ca^{2+} of 570 nM may also underestimate changes in $[Ca^{2+}]_{mt}$, and low affinity indicators may be required to give a more accurate reflection of changing $[Ca^{2+}]_{mt}$.

6 Measuring changes in [ATP]

In general, if you are interested in measuring changes in mitochondrial function, you are interested in cell metabolism, and the key end-point of much of this is ATP. There are innumerable physiological and pathophysiological experimental questions where a key question is: what is happening to [ATP] or even more importantly, the ATP/ADP.P_i ratio? At present, there are no direct fluorescent indicators for ATP or ADP. A fluorescent probe for P_i might be useful, but the only probe described has a K_D which is too low to be useful to follow changes during cell metabolism. To our knowledge, two solutions have been introduced to study [ATP] in cells: the first involves the microinjection or the transfection of cells with luciferase. The second is highly indirect, and involves the measurement of $[Mg^{2+}]_i$.

6.1 Luciferase

Luciferase generates a chemiluminescent signal in the presence of ATP and luciferin. Cooled photon counting photomultipliers placed immediately over the stage of a microscope without any intervening optics were used by Cobbold's group (43) to measure ATP within single cardiomyocytes microinjected with luciferase. This was a substantial feat and is not to be undertaken lightly. More recently, it seems more attractive to persuade the cells to manufacture the luciferase themselves, especially as constructs are readily available. In general, chemiluminescence signals are very dim, and so hard to image effectively at the level of the single cell. By the far the bulk of work in this area (using aequorin chemiluminescence as well) involves recording signals from populations of cells to yield averaged signals from the population. Kennedy *et al.* (44) have recently described measurements of luminescence of luciferase engineered to signal ATP levels from different compartments. They have also tried to image luciferase luminescence, using an intensified CCD system (Photek). Although integration times are necessarily long, this approach does have the merit of defining cell to cell variations in responses.

6.2 $[Mg^{2+}]_i$

This may sound obscure, but it relies on the differential affinities of ATP and ADP for Mg^{2+}: that of ADP is about tenfold lower than that of ATP, and so ATP hydrolysis tends to liberate Mg^{2+}. The approach is not only indirect but is a minefield of potentially misleading observations, and so approach with care! First, no $[Mg^{2+}]_i$-sensitive dye is available that is sensitive only to Mg^{2+}. The affinities of these dyes for Mg^{2+} lie in the mM (i.e. physiological) range and in the micromolar range for Ca^{2+} (i.e. rather high to be physiological). For example Magnesium Green (MgG) has a K_D for Ca^{2+} of 4.7 μM, while the K_D for Mg^{2+} is 0.9 mM.

However, many conditions in which ATP is depleted are also associated with a rise in $[Ca^{2+}]_i$. In our own study of ATP depletion in cardiomyocytes, this problem was avoided firstly by making simultaneous measurements of $[Ca^{2+}]_i$ and $[Mg^{2+}]_i$—the clear divergence between the signals made it clear that the $[Mg^{2+}]_i$ indicator was not simply measuring $[Ca^{2+}]_i$ (see below). It was also possible to carry out the same experiments in the absence of external Ca^{2+} and with $[Ca^{2+}]_i$ buffered, so that $[Ca^{2+}]_i$ signals were completely flat, while the $[Mg^{2+}]_i$ signals were unchanged, but this will obviously not be possible if the subject of study is a Ca^{2+}-dependent process—cell death, for example.

The dyes MgG and Mag-Indo have both been used in this way, and have clearly provided very useful data on the mechanisms of ATP depletion in cardiomyocytes, although a similar approach has proved much more difficult in other cell types (unpublished observations).

The answer at present seems to be a plea to the chemists for a useful fluorescence probe to measure ATP, ADP, AMP, or P_i. All would be invaluable.

7 Cytochrome absorptance measurements

7.1 Theoretical basis

The energy stored in the chemical bond in metabolic substrates is harnessed and used efficiently by transferring the electrons to molecular oxygen via a series of steps, the so-called electron transfer chain. This enzyme cascade is localized in the mitochondrial inner membrane, and consists of several different types of molecules which include the cytochromes. As indicated even by their name, cytochromes absorb light with characteristic absorptance spectra that have fairly broad peaks. Since the pioneering work of Keilin (45) and later by Chance (for review see ref. 46) it has been known that the wavelengths of these peaks shift slightly when their redox state changes, and the absorptance difference spectra (i.e. the difference between spectra under oxidized and reduced conditions) have very distinct features that are identical in all organisms. This is an area which still has not been greatly exploited despite the pioneering work of Chance and colleagues in the 1950s and 60s. Nevertheless, it provides another approach to our attempts to follow changes in mitochondrial status within tissues, and so warrants some careful thought.

Based on their optical properties, the cytochromes were originally classified as '*a*', '*b*', and '*c*'. Each cytochrome has three clear maxima, called Soret bands and labelled α, β, and γ. Later it became apparent that these optical species actually consist of more than three cytochromes (named a, a_3, c_1, and c; b consists of many species). CN^-, an inhibitor of mitochondrial respiration, replaces O_2 from the last cytochrome in the chain, cytochrome a_3. This causes a conformational shift similar to the oxidized state. Thus whereas all other cytochromes are fully reduced as no electrons can be transferred to O_2, cytochrome a_3 appears to be in the oxidized state.

7.2 Methods and limitations

Measurement of changes in cytochrome absorptance requires special instrumentation. The absorptance of mitochondrial pigments is low, apparently precluding measurements in single cells. However cell suspensions can be studied using a stirred, thermostatically controlled cuvette, and small tissue samples (hippocampal slices, heart papillary muscle, carotid body, or the blowfly compound eye, a preparation very rich in mitochondria) can be mounted in a perfusion chamber on a microscope stage. The illumination light should be provided by a stable light source with homogeneous spectral output, for example a halogen lamp with stabilized power supply. The spectrograph could either be a scanning spectrograph with an exit slit and photomultiplier, or a stationary spectrograph fitted with a detector array (e.g. CCD or photodiode array) which allows recording the full spectrum instantaneously. The spectrograph grating can be quite elementary and inexpensive as there is no need for high spectral resolution; 1–2 nm is sufficient since the peaks in the absorptance difference spectra have a full width at half maximum of about 10–15 nm. When very small changes in absorptance are to be expected, the type of detector becomes important. The detector should have low dark noise and high dynamic range (at least 12-bit, with as large a full well depth as possible). In addition the faster the sampling rate, the more one can average spectra without affecting time resolution, to increase sensitivity to small changes.

One of the problems with absorptance spectroscopy is non-specific absorption changes due to changes in the light diffusion properties of the tissue. For example hypoxia may cause cell swelling, affecting the reflection on water–membrane interfaces. This can be overcome by using a ratiometric approach, similar as in the use of fura-2: the changes at the peak absorptance are compared to changes at wavelengths where cytochrome absorptance is independent of redox state. In principle one could use photomultipliers or photodiodes and band pass filters. By choosing appropriate wavelengths and ratioing the outputs of two detectors, the signal obtained will be more robust against specific changes of absorption, but limited to one cytochrome only. Alternatively, a spectrograph provides full spectra which can be used to ratio the peak absorptance against non-specific absorptance. These ratiometric approaches depend on reliable isosbestic points. Unfortunately, flavoproteins and co-enzyme Q also absorb light, albeit with a much smaller amplitude and over much larger ranges of the spectrum. A perhaps more reliable approach may be to use known cytochrome difference spectra, and to deconvolve the obtained spectra to extract the changes in redox state of the individual cytochromes. However these techniques are not common practise.

With the cytochromes the same caveat applies as with NADH and flavoprotein (see Section 3.3), namely that knowledge of changes in redox state does not *per se* indicate whether there is an increase or a decrease in electron flux, since the redox state depends on the sum of reduction and oxidation rates.

7.3 Perspectives

Recently a new type of spectrograph has come on the market. Most spectrographs use spherical mirrors losing all spatial information along the entrance slit. In contrast, imaging spectrographs (e.g. the MS127i from L.O.T.—Oriel or the Triax series from Jobin Yvon & Glen Spectra) use toroidal mirrors, which image the entrance slit onto the grating. Light that enters at the top of the slit will be imaged on the top of the grating, and will exit the spectrograph at the top. If a two-dimensional CCD chip is used as detector, the spectral properties of the light entering the spectrograph at the top of the slit is recorded on the top rows of the CCD array. Thus in principle using a 1024 × 128 CCD array one could obtain the spectral properties of 128 points along a vertical line. In practice at least five different areas can be sampled simultaneously.

Imaging spectrographs were designed to be used with bifurcating light guides, that allow simultaneous measurement of more than one sample. However when the entrance slit of the imaging spectrograph is placed in the output focal plane of a microscope, changes in the spectral properties along a line in tissue samples can be studied. For example it should become possible to resolve different adjacent layers in hippocampal slices. The position of the line can be visualized with the same CCD detector head, by rotating the grating such that the 0th order maximum is projected onto the CCD detector. After selecting the desired area, the grating is rotated back to project the 1st order maximum onto the detector. The most convenient way to recalibrate the spectrograph is to acquire the spectrum of overhead fluorescent tubing, and to use the most prominent Hg, Ne, and Ar lines (in nm: triplet at 365; doublet at 405/408; 436, 546, doublet at 577/579; 808; 811; triplet at 871; 1014).

8 Simultaneous measurements of mitochondrial and other cellular parameters

In many applications it would be useful to be able to record mitochondrial and other cellular parameters simultaneously using different fluorescent probes. The selection of the dyes used for such measurements is dependent upon the excitation and emission spectra of these probes, and it is always essential to ensure that the signals are clearly discrete and that there is no cross-talk or energy transfer between signals.

A simple option is to use fluorescence probes which have different excitation spectra and similar emission peaks. In such cases, a chopper, spinning filter wheel, monochromator, or other mechanism for switching between excitation wavelengths can be used. The dichroic mirror must allow efficient reflection of all excitation maxima. Alternatively, dyes may be chosen with the same, or similar, excitation spectra, but with distinct emission peaks, in which case two detectors (photomultiplier tubes—to use two cameras is expensive and they are difficult to align) together with a secondary dichroic reflector are required to measure the two emission signals. This approach is fairly routine using confocal microscopy, one of the great merits of a confocal imaging system.

8.1 Simultaneous monitoring of $\Delta\psi_m$ and $[Ca^{2+}]_i$

Because the mitochondria represent one of the main cellular $[Ca^{2+}]_i$ homeostatic systems, and regulation of $[Ca^{2+}]_i$ is critically dependent on mitochondrial function, examination of the temporal relationships between changes in $[Ca^{2+}]_i$ and $\Delta\psi_m$ is of major interest. For simultaneous monitoring of $\Delta\psi_m$ and $[Ca^{2+}]_i$ it is convenient to use the Ca^{2+} indicator fura-2 or its low affinity derivative, fura-2 FF ($K_D \sim 35$ μM), together with the potentiometric dye Rh123 or TMRE. The excitation spectra of fura-2 (340/380 nm) and Rh123 (490 nm) are sufficiently distinct to allow separate excitation without contamination, while the emission spectra ($\sim$ 510 nm for fura-2 and $\sim$ 530 nm for Rh123) are sufficiently similar to allow measurement of both fluorescence signals with a single detector. Such measurements have been effectively used in experiments with cultured astrocytes, cardiomyocytes, adrenal chromaffin cells, dissociated and cultured central mammalian neurons (please see Plate 5A).

Our protocols involve conventional loading with fura-2 AM, while the Rh123 is added during the final 15 minutes of loading time.

It is also possible to use dyes like fluo-3 or Calcium Green to measure $[Ca^{2+}]_i$ with TMRE or TMRM to measure $\Delta\psi_m$ on a dual emission photometric system or confocal imaging system. Even though the excitation peak for TMRM and TMRE are at about 540 nm, they are both excited efficiently at ~ 490 nm or with the 488 nm laser required to excite the $[Ca^{2+}]_i$ indicators, and the emission spectra can be measured separately on two channels at 505–530 nm (fluo-3) and > 570 nm (TMRE) separated by a dichroic mirror at ~ 550 nm.

No doubt other combinations are practical.

A technical point: Most dichroic mirrors only reflect efficiently over a relatively narrow wavelength range. If you wish to use fura-2 and Rh123, it is optimal to acquire a dichroic mirror that has an extra coating that reflects efficiently over the full range from 320 nm or so up to 510 nm, and transmits > 515 nm (Chroma call this the 'Tsien special').

8.2 Simultaneous recording of $\Delta\psi_m$ and autofluorescence

Measurements of changes in NADH autofluorescence are more readily interpreted when coupled with measurements of $\Delta\psi_m$, and these two variables can be measured simultaneously. For combined records of mitochondrial potential and NADH autofluorescence a potentiometric indicator—Rh123, one channel of JC-1 or TMRE may be used in combination with autofluorescence measurements. Because both excitation and emission spectra of NADH and Rh123 etc. are distinct, a double reflectance band dichroic mirror is required in the epifluorescence light path to allow illumination with light at 360 nm for excitation of NADH and 490 for JC-1 with transmission bands from ~ 400–480 nm and then above 510 nm to allow simultaneous collection of both the NADH autofluorescence (with a peak at 450 nm) and the Rh123 signal (with a peak at 530 nm). The fluorescence signal from two detectors is then separated by dichroic mirror at 510 nm. Again, this is possible on a UV confocal system equipped with appropriate filters on collection channels (please see Plates 5C and 5D).

8.3 Simultaneous recording of $[Ca^{2+}]_{cyt}$ and $[Ca^{2+}]_{mt}$

Recently we demonstrated in experiments with cultured astrocytes (4) that, as rhod-2 was partitioned between the cytosol and mitochondria, the remaining cytosolic signal in the nucleus (a mitochondrion-free zone) could be effectively used to monitor changes in $[Ca^{2+}]_{cyt}$ while the signal over mitochondria could be isolated either using a confocal imaging system or using a cooled CCD system with very flat cells. Because in most cells Ca^{2+} equilibrates rapidly between the cytosol and the nucleus, the changes in nuclear $[Ca^{2+}]$ is very similar to the changes in $[Ca^{2+}]_i$ and the fluorescence signal from the nucleus may be interpreted in terms of changes in $[Ca^{2+}]_{cyt}$ (please see Plate 4).

Rhod-2 may be also used in combination with other more pancellular distributed Ca^{2+} indicators to enable concurrent measurements of mitochondrial and cytosolic [Ca^{2+}]. In chromaffin cells concurrent monitoring of [Ca^{2+}]$_{cyt}$ and [Ca^{2+}]$_{mt}$ has been performed using rhod-2 and dyes such as Calcium Green or fluo-3 (30). To this end a filter cube with dual band excitation (490 and 560 nm) and dual band emission (530 and 650 nm) (XF-52; Omega Optical) may be most appropriate, and detection at 525 and 580 nm was accomplished using two photomultiplier tubes with additional interference filters separated by 540 nm dichroic mirror (30). The same combination may be used effectively on a confocal system (please see Plate 6B). The beauty of this approach is that measurements can be made from the mitochondrial and cytosolic indicators in precisely the same volume of the cell and from signals acquired at precisely the same time point, and this raises the possibility of precise measurements of the temporal relationship between focal cytosolic [Ca^{2+}] and local mitochondrial uptake.

Many other combinations are possible: simultaneous measurements of Magnesium Green ([ATP]) and [Ca^{2+}]$_{cyt}$, etc. We have yet to find a satisfactory way to measure mitochondrial [Ca^{2+}] and $\Delta\psi_m$ simultaneously, as there has been too much cross-talk or spectral overlap between all the combinations of dyes that we have tried to date.

References

1. McCormack, J. G., Halestrap, A. P., and Denton, R. M. (1990). *Physiol. Rev.*, **70**, 391.
2. Thayer, S. A. and Miller, R. J. (1990). *J. Physiol. (Lond.)*, **425**, 85.
3. Jouaville, L. S., Ichas, F., Holmuhamedov, E. L., Camacho, P., and Lechleiter, J. D. (1995). *Nature*, **377**, 438.
4. Boitier, E., Rea, R., and Duchen, M. R. (1999). *J. Cell Biol.*, **145**, 795.
5. Duchen, M. R. (1999). *J. Physiol. (Lond.)*, **516**, 1.
6. Stout, A., Raphael, H., Kanterewicz, B., Klann, E., and Reynolds, I. (1998). *Nature Neurosci.*, **1**, 366.
7. Rizzuto, R., Bastianutto, C., Brini, M., Murgia, M., and Pozzan, T. (1994). *J. Cell Biol.*, **126**, 1183.
8. Bernardi, P., Scorrano, L., Colonna, R., Petronilli, V., and Di Lisa, F. (1999). *Eur. J. Biochem.*, **264**, 687.
9. Crompton, M. (1999). *Biochem. J.*, **341**, 233.
10. Noji, H., Yasuda, R., Yoshida, M., and Kinosita, K. J. (1997). *Nature*, **386**, 299.
11. Gunter, K. K. and Gunter, T. E. (1994). *J. Bioenerg. Biomembr.*, **26**, 471.
12. Duchen, M., McGuinness, O., Brown, L., and Crompton, M. (1993). *Cardiovasc. Res.*, **27**, 1790.
13. Duchen, M. R. and Biscoe, T. J. (1992). *J. Physiol. (Lond.)*, **450**, 33.
14. Leyssens, A., Nowicky, A., Patterson, L., Crompton, M., and Duchen, M. (1996). *J. Physiol. (Lond.)*, **496**, 111.
15. Chance, B., Schoener, B., Oshino, R., Itshak, F., and Nakase, Y. (1979). *J. Biol. Chem.*, **254**, 4764.
16. Chance, B. and Baltscheffsky, H. (1958). *J. Biol. Chem.*, **233**, 736.
17. Mojet, M. H. (1992). Ph.D. Thesis, University of Groningen, Groningen, The Netherlands.
18. Duchen, M. (1992). *Biochem. J.*, **283**, 41.
19. Mitchell, P. and Moyle, J. (1965). *Nature*, **208**, 1205.
20. Loew, L. M., Tuft, R. A., Carrington, W., and Fay, F. S. (1993). *Biophys. J.*, **65**, 2396.

21. Di Lisa, F., Blank, P. S., Colonna, R., Gambassi, G., Silverman, H. S., Stern, M. D., *et al.* (1995). *J. Physiol. (Lond.)*, **486**, 1.
22. Chinopoulos, C., Tretter, L., and Adam-Vizi, V. (1999). *J. Neurochem.*, **73**, 220.
23. Reers, M., Smith, T. W., and Chen, L. B. (1991). *Biochemistry*, **30**, 4480.
24. Smiley, S. T., Reers, M., Mottola-Hartshorn, C., Lin, M., Chen, A., Smith, T. W., *et al.* (1991). *Proc. Natl. Acad. Sci. USA*, **88**, 3671.
25. Fink, C., Morgan, F., and Loew, L. W. (1998). *Biophys. J.*, **75**, 1648.
26. Nowicky, A. and Duchen, M. (1998). *J. Physiol. (Lond.)*, **507**, 131.
27. Rottenberg, H. and Wu, S. L. (1998). *Biochim. Biophys. Acta*, **1404**, 393.
28. Emaus, R. K., Grunwald, R., and Lemasters, J. J. (1986). *Biochim. Biophys. Acta*, **850**, 436.
29. Hüser, J., Rechenmacher, C. E., and Blatter, L. A. (1998). *Biophys. J.*, **74**, 2129.
30. Miyata, H., Silverman, H. S., Sollott, S. J., Lakatta, E. G., Stern, M. D., and Hansford, R. G. (1991). *Am. J. Physiol.*, **261**, H1123.
31. Babcock, D. F., Herrington, J., Goodwin, P. C., Park, Y. B., and Hille, B. (1997). *J. Cell Biol.*, **136**, 833.
32. Drummond, R. M. and Tuft, R. A. (1999). *J. Physiol. (Lond.)*, **516**, 139.
33. Csordás, G., Thomas, A. P., and Hajnóczky, G. (1999). *EMBO J.*, **18**, 96.
34. Delcamp, T. J., Dales, C., Ralenkotter, L., Cole, P. S., and Hadley, R. W. (1998). *Am. J. Physiol.*, **275**, H484.
35. Griffiths, E. J., Stern, M. D., and Silverman, H. S. (1997). *Am. J. Physiol.*, **273**, C37.
36. Gavin, C. E., Gunter, K. K., and Gunter, T. E. (1999). *Neurotoxicology*, **20**, 445.
37. Robb-Gaspers, L. D., Burnett, P., Rutter, G. A., Denton, R. M., Rizzuto, R., and Thomas, A. P. (1998). *EMBO J.*, **17**, 4987.
38. Trollinger, D. R., Cascio, W. E., and Lemasters, J. J. (1997). *Biochem. Biophys. Res. Commun.*, **236**, 738.
39. Hajnóczky, G., Robb-Gaspers, L. D., Seitz, M. B., and Thomas, A. P. (1995). *Cell*, **82**, 415.
40. Simpson, P. B. and Russell, J. T. (1998). *J. Physiol. (Lond.)*, **508**, 413.
41. Peng, T. I., Jou, M. J., Sheu, S. S., and Greenamyre, J. T. (1998). *Exp. Neurol.*, **149**, 1.
42. Matlib, M. A., Zhou, Z., Knight, S., Ahmed, S., Choi, K. M., Krause-Bauer, J., *et al.* (1998). *J. Biol. Chem.*, **273**, 10223.
43. Bowers, K. C., Allshire, A. P., and Cobbold, P. H. (1992). *J. Mol. Cell. Cardiol.*, **24**, 213.
44. Kennedy, H. J., Pouli, A. E., Ainscow, E. K., Jouaville, L. S., Rizzuto, R., and Rutter, G. A. (1999). *J. Biol. Chem.*, **274**, 13281.
45. Keilin, D. (1966). *The history of cell respiration and cytochrome*. Cambridge University Press, Cambridge, UK.
46. Chance, B. (1965). *J. Gen. Physiol.*, **49**, Suppl: 163.
47. Nicholls, D. G. and Ferguson, S. J. (1992). *Bioenergetics 2*. Pub Academic Press, London, UK.
48. Pearce, R. and Duchen, M. (1995). *J. Physiol. (Lond.)*, **483**, 407.
49. Crompton, M., Ellinger, H., and Costi, A. (1988). *Biochem. J.*, **255**, 357.
50. Crompton, M. and Andreeva, L. (1994). *Biochem. J.*, **302**, 181.
51. Duchen, M. R., Leyssens, A., and Crompton, M. (1998). *J. Cell Biol.*, **142**, 975.
52. Cox, D. A., Conforti, L., Sperelakis, N. and Matlib, M. A. (1993). *J. Cardiovasc. Phamacol.*, **21**, 595.
53. Duchen, M., Smith, P., and Ashcroft, F. (1993). *Biochem. J.*, **294**, 35.

Appendix

1 The mitochondriac's basic pharmacopoeia

A clear understanding of the basic biochemical pharmacology of the mitochondrion and of the actions of these reagents is crucial in the approach to studying mitochondria, as it is possible to manipulate each aspect of mitochondrial function in a controlled way to assess the consequences for the variable under study. For greater detail, please refer to Nicholls and Ferguson (47).

2 Inhibitors of mitochondrial respiration

2.1 Complex I

Rotenone (Sigma) is the classic inhibitor at complex I. It is soluble in alcohol and is usually used at concentrations of ~ 1 μM. It tends to deteriorate in storage and may develop a yellow discoloration which can cause artefacts in fluorescence measurements.

Complex I is also blocked by a number of *barbiturates*. The classic is amylobarbitone (~ 1 mM), but a number of other barbiturates are very potent inhibitors at this site (see ref. 48).

2.2 Complex III

Antimycin A and myxothiazol (Sigma). These two agents both inhibit the bc_1 complex, i.e. at complex III. They are routinely used at low micromolar concentrations, for example we use antimycin A at 1 μM. There are subtle differences in the site of action of each of these agents which are beyond the scope of this essay. The biophysics is explained in some depth by Nicholls and Ferguson (47).

2.3 Complex IV

Cyanide (CN^-, Merck) the classic poison of the murder mystery, is also the classic inhibitor at complex IV, cytochrome *c* oxidase, the final step of the mitochondrial respiratory chain, where it inhibits the reduction of oxygen. It is usually used a concentrations of 1–2 mM. This is supramaximal, but takes into account the volatility of the drug—it is likely that the effective concentration reaching the cells will be significantly less. In solution it is quite alkaline, and so care must be taken to adjust the pH. At higher concentrations, CN^- also inhibits the cytosolic CuZn superoxide dismutase (SOD).

Azide (Sigma). Sodium azide can also be used as an inhibitor of complex IV, and seems to be used at low millimolar concentrations in cells. The use of azide in cells may be more problematic than cyanide however because cellular actions, other than inhibition of complex IV, are increasingly being reported for this compound.

Anoxia. Perhaps the 'cleanest' way to stop mitochondrial respiration is to remove oxygen. This is less easy than it might sound: bear in mind that mitochondrial respiration in most cells is not significantly compromised until the PO_2 falls below 1 mmHg (in most cells under most conditions), while the PO_2 in ambient room air is ~ 150 mmHg, and that in most tissues is ~ 20–40 mmHg. In order to make cells anoxic (and it is necessary here to be pedantic about the differences between *an*oxia and *hyp*oxia) we have bubbled perfusates with argon or nitrogen (argon is heavier and so tends to form a layer over the meniscus in a dish of cells). After purging as much of the oxygen as possible by equilibration with argon, it is then necessary to use oxygen impermeant tubing throughout the perfusion system. Using a trough-shaped bath, oxygen pressures below 1 mmHg are achievable, but it may be necessary either to use a closed perfusion chamber filled with argon or nitrogen or to add about 500 μM sodium dithionite ($Na_2S_2O_4$) to scavenge any contaminating oxygen. Do not add the dithionite before purging the oxygen from the saline as it will simply be oxidized and you will have to use a high concentration (which is then very acidic) to lower PO_2. It is always advisable if possible in such experiments also to use a PO_2 electrode close to the cells to ensure that the PO_2 really is falling as low as you think it is.

2.4 Uncouplers

Mitochondrial uncouplers operate as proton ionophores, shuttling protons across the inner mitochondrial membrane. Once exposed to an uncoupler, mitochondrial potential collapses and respiration becomes uncontrolled. The usual agents used are FCCP (Sigma) (used at a final concentration of ~ 1 μM, stock in DMSO) or CCCP (used at ~ 5–10 μM). Dinitrophenol (DNP) is also sometimes used. All act very quickly—within a few seconds. FCCP will certainly wash out and can apparently be fully reversible. Remember also that these agents act at all membranes and dissipate all proton gradients—across the plasma membrane, ER/SR membrane, and across vesicular membranes, and therefore cannot be selective for mitochondria. It is a common usage to assume that if application of FCCP raises $[Ca^{2+}]_i$ this necessarily reflects mitochondrial releasable Ca^{2+}—while this is a possible interpretation, FCCP may deplete ATP, it causes an acidification of pH_i, and may have other actions at the plasma membrane.

2.5 The F_1F_0-ATP(synth)ase

Oligomycin (Sigma) is the major inhibitor of this enzyme that phosphorylates ADP to produce ATP. It derives the necessary energy from the downhill movement of protons through the F_0 component—essentially a proton channel. Oligomycin is most commonly used as a mixture of oligomycins A, B, and C, at concentrations of 2.5 μg/ml. Oligomycin will dissolve in ethanol as a stock. Its action appears to be completely irreversible, and it will prevent both the synthesis of ATP and its consumption by the enzyme in its reverse mode.

2.6 Electron donors

TMPD/ascorbate (Sigma). Tetramethyl-*p*-phenylenediamine (TMPD) is a membrane-permeant redox dye which can donate electrons to cytochrome *c*, thus enabling the first three complexes of the mitochondrial respiratory chain to be effectively bypassed. The combination of TMPD and ascorbate ensures that a cycle of oxidation and reduction can occur, so that maximal donation of electrons to cytochrome *c* is achieved. These compounds are widely accepted as substrates for complex IV respiration, and the concentrations routinely used are 20 μg/ml of TMPD and 1 mM ascorbate. Care has to be taken not to use higher concentrations of ascorbate, because other cellular processes then become affected. For example, 3 mM ascorbate inhibits NMDA receptor fluxes in neurons. TMPD is oxidized to a blue product, which may interfere with optical measurements and we have seen curious apparently artefactual changes in Indo-1 and in autofluorescence signal using this reagent, and so care must be taken in making and interpreting such measurements.

2.7 The adenine nucleotide translocase

Atractyloside (Sigma) and **bongkrekate** (Calbiochem). These drugs both inhibit the adenine nucleotide translocase (ANT). They therefore prevent the export of mitochondrial ATP but will also cause depletion of intramitochondrial ADP, preventing the import of substrate for phosphorylation. It is now clear that the ANT forms an integral part of the protein complex that forms the permeability transition pore (PTP, see below), and it was the action of these two drugs that first drew attention to this role. Atractyloside will open the PTP while bongkrekate stabilizes the closed state. Bongkrekate crosses membranes readily, but accumulates in mitochondria in response to $\Delta\psi_m$. It has been used at 10–100 μM in intact cells, and at tenfold lower concentrations in permeabilized cells. Atractyloside is probably not very membrane-permeant—the literature is confusing, as some have applied it to intact cells and seen effects, while other groups have gone to some lengths to introduce it into cells with microinjection or lipofection. We have never seen any significant effect when applied to intact cells.

2.8 The permeability transition pore

Cyclosporin A (CsA) (Sigma) closes the PTP (49). In cells, CsA may show a bell-shaped dose–response curve in terms of cell *protection* from anoxia/reperfusion injury, with a peak at about 300 nM. It must be stressed that CsA acts by binding to cyclophilins, a family of proteins only one of which, cypD, is involved in regulation of the permeability transition pore. Cyclophilins have many roles in cells, which include the regulation of the calcium-dependent protein phosphatase calcineurin. An analogue of CsA, **methyl-valine CsA** (mvCsA; by written request to Novartis) has been introduced which forms a complex with the cytosolic cyclophilin CypA which does not modulate calcineurin, but still binds all cyclophilins (this is often cited as being more selective for the PTP than CsA, which is not strictly correct). Thus, while CsA will suppress PTP opening, this is not really a reliable single guarantor that a mitochondrial depolarization truly reflects PTP. The drug FK506 (Calbiochem) binds to FK binding proteins which also act at calcineurin but have no effect on the PTP, and so this may be a useful control. CsA and mvCsA should be effective at concentrations of about 100–500 nM. A pre-incubation time of about 30 minutes seems useful.

2.9 The calcium uniporter

The classic inhibitor of the uniporter is **ruthenium red** (RuR) (Sigma). This is an impure compound which has many other actions, blocking a variety of other classes of ion channel (mostly calcium-permeant channels in both the plasma membrane and in sarco- and endoplasmic membranes). From its structure, it should be expected to be membrane-impermeant. There are a number of reports of effects of ruthenium red on cardiomyocyte function that assume permeation. More recently, **Ru360** (Calbiochem) has been synthesized as a membrane-permeant analogue of RuR. At the time of writing, the data on the properties of this agent are limited (see ref. 42), but in our hands, preliminary data suggest that 10–100 μM concentrations may have other actions affecting cellular calcium handling (D. Jacobson and E. Boitier, unpublished observations), whereas concentrations below this had no effect on mitochondrial calcium uptake.

Crompton and Andreeva (50) synthesized a range of cobalt-based compounds as inhibitors of the uniporter. Of these, the most effective was diamino pentane pentammic acid (DAPPAC, see also ref. 51). This is not commercially available, as yet, but is a potent inhibitor of the uniporter and does not appear to block SR Ca^{2+} channels. It is not membrane-permeant and must be microinjected into cells.

2.10 The Na^+/Ca^{2+} exchanger

The Na^+/Ca^{2+} exchanger is a major route for the removal of intramitochondrial Ca^{2+} and allows the slow re-equilibration of Ca^{2+} pools following mitochondrial Ca^{2+} accumulation. The major agent used now is CGP37157 (by written request to Ciba Geigy Pharmaceuticals, Switzerland). Cox *et al.* (52), reported an IC50 of < 1 μM and showed that it was reasonably selective in intact cells at concentrations < 10 μM but most seem to use it at 25 μM.

Chapter 6

Using low-affinity fluorescent calcium indicators and chelators for monitoring and manipulating free [Ca^{2+}] in the endoplasmic reticulum

Aldebaran M. Hofer
Harvard Medical School, Brigham and Women's Hospital and the West Roxbury VAMC, West Roxbury, MA 02132, USA.

1 Introduction

To date, the intracellular calcium signal has been the most intensively studied of the various second messenger pathways. Without question, the major factor in the extraordinary success of this field has been the widespread availability of sensitive, reliable, high-affinity fluorescent Ca^{2+} indicators, such as those originally introduced by Roger Y. Tsien and colleagues (1, 2). The parallel development of membrane-permeant AM-ester forms of dyes that could be easily loaded into cells further simplified the technique, making the use of these probes accessible to an even larger community of researchers.

Of central importance to Ca^{2+} signalling events in the cytoplasm are subcellular Ca^{2+} storage compartments such as the endoplasmic reticulum (ER), which serve as essential sources and sinks of Ca^{2+} following agonist-induced Ca^{2+} mobilization (3). Although it has been possible to deduce many of the Ca^{2+} transport properties of internal stores indirectly by measuring [Ca^{2+}] within the cytosol, not all aspects of ER Ca^{2+} handling can be readily predicted in this manner (4). Moreover, there is increasing evidence that calcium ions may fulfil specialized signalling roles in the lumen of the ER, as well as in other cellular compartments. Thus, complete understanding of cellular Ca^{2+} homeostasis and the potential regulatory actions of Ca^{2+} ions within organelles call for direct measurement of the cation within subcellular compartments. The purpose of this chapter is to describe various practical aspects of using compartmentalized low-affinity fluorescent indicators for monitoring [Ca^{2+}] within the ER ([Ca^{2+}]$_{ER}$). In addition, some attention will be devoted to recently described techniques for manipulating free luminal [Ca^{2+}] using the membrane-permeant Ca^{2+} chelator, tetrakis-(2-pyridylmethyl)ethylenediamine (TPEN).

1.1 Basic principles

Organellar trapping of AM-ester derivatives has long been a source of frustration for investigators seeking information about [Ca^{2+}] dynamics exclusively in the cytoplasm. Terasaki and Sardet were among the first to recognize the potential for exploiting dye compartmentalization, having noted the accumulation of fluo-3 AM into the cortical ER of sea urchin eggs (5). Other investigators attempted to use trapped fura-2 for the purpose of imaging subcellular Ca^{2+} release sites in mammalian cells (6–8). In retrospect, it is likely that these high affinity indicators (K_d for Ca^{2+} for fluo-3 and fura-2 = 309 nM and 229 nM, respectively) were largely saturated in the ER, an organelle now known to maintain a free resting [Ca^{2+}] of several hundred μM or more. Nevertheless, dynamic changes attributed to agonist-induced Ca^{2+} fluxes could be observed under special conditions, such as when stores had already been partially depleted of their Ca^{2+}.

The use of compartmentalized low-affinity Ca^{2+} indicators such as mag-fura-2 (K_d for Ca^{2+} = 53 μM) (9) and mag-Indo-1 (K_d for Ca^{2+} = 32 μM) however, allowed direct monitoring of $[Ca^{2+}]_{ER}$ changes under more ‘physiological’ circumstances (10–12). The procedures involved in the use of this approach are quite elementary. Cells are loaded with the AM form of the low-affinity probe, resulting in dye accumulation in the cytoplasm and organelles. Fluorophore remaining in the cytosol is eliminated, either by dialysis with a patch pipette or by permeabilization of the plasma membrane. Fluorescence from residual organelle-trapped dye is then monitored as is conventionally performed for the cytoplasm. More recently, it has become apparent that some cell types (fibroblasts, astrocytes, smooth muscle) exclude dye from the cytoplasm under certain loading conditions, resulting in preferential accumulation in organelles (13–15). In these cell types, sensitive measurements of free [Ca^{2+}] changes in internal stores can be performed in intact cells, as described below.

1.2 Mechanism of dye uptake

AM ester derivatives have been shown to accumulate in a variety of subcellular structures including the ER, nuclear envelope (technically a subcompartment of the ER), secretory granules, and mitochondria. The selectivity of compartmentalized dye techniques for the measurement of [Ca^{2+}] within agonist-sensitive stores is therefore based largely on the fact that the particular indicators used for this approach have low affinities for Ca^{2+}. In principle, [Ca^{2+}] fluctuations occurring in organelles such as mitochondria, which maintain a resting free [Ca^{2+}] slightly higher than the cytoplasm, are below the limit of detection of these probes. It should be noted that dye in these ‘silent’ compartments nevertheless contributes to the overall fluorescence measured from the cell (16) unless sophisticated optical techniques are used to resolve signals from individual, identified organelles (15, 17).

Exactly how indicators loaded as AM esters come to reside in organelles and why different cell types vary so much in their dye handling characteristics is unknown. In cells that have been permeabilized, internal stores readily load with the AM ester forms of dyes, but not the free acid (A. M. Hofer, unpublished results). Mammalian carboxylesterases that can use AM esters as substrates are part of a well-characterized multigene family. Isoforms specific to the ER, mitochondria, and the cytoplasm have been identified (18). Thus a likely scenario is that AM derivatives diffuse, or are possibly transported (for example by membrane transporters belonging to the p-glycoprotein family) (19), into internal stores where they are cleaved by luminal esterases. Evidently, this cleavage can be extremely efficient. BAPTA-AM added acutely to intact BHK-21 cells was observed to rapidly (within minutes) lower the luminal $[Ca^{2+}]$ (implying disruption of the ester linkage), as reported by mag-fura-2 within the stores (14). The decrease in free $[Ca^{2+}]_{ER}$ was sufficient to evoke robust capacitative Ca^{2+} entry (see below on the intentional use of TPEN for this purpose).

The nature and number of Ca^{2+} stores labelled with dye (e.g. $InsP_3$-dependent versus $InsP_3$-independent pools) is largely a function of cell type. For example, in RBL-1 cells, compartmentalized mag-fura-2 reports $[Ca^{2+}]$ changes in at least two pharmacologically distinct compartments. The first can be released to the same extent by either the selective SERCA-inhibitor, thapsigargin or by a maximal dose of $InsP_3$. In these cells, residual Ca^{2+} is retained in a small pool (or pools) sensitive to the Ca^{2+}/H^{+} exchanging ionophore, ionomycin. In contrast, in BHK-21 fibroblasts, all releasable Ca^{2+} (as reported by the dye) is localized to one functionally defined compartment sensitive to thapsigargin (20). No additional Ca^{2+} release can be detected following ionomycin treatment. $InsP_3$ releases a large fraction of the stored Ca^{2+} (about 85%) in this pool. Detailed analysis of the dye distribution in these cells revealed that *at least* 88% of the compartmentalized probe was localized to the thapsigargin-releasable internal store (16). The possibility, however, that remaining dye (up to 12%) was contained in organelles that maintain a low resting $[Ca^{2+}]$ (such as mitochondria) could not be excluded.

1.3 Choice of indicator

A wide variety of low-affinity fluorescent Ca^{2+} indicators are now commercially available, and many of these have been used successfully for measuring $[Ca^{2+}]_{ER}$ (*Table 1*). These probes have varying spectral characteristics, different affinities for Ca^{2+} and Mg^{2+}, and other properties that may recommend them for a particular application. Practically speaking, the selection of indicator is usually dictated by the type of measuring system available to the experimenter. For example, if the laboratory is equipped for ratiometric imaging with fura-2, then mag-fura-2 can be used with little or no modification of the existing set-up. Mag-fura-2 (also known as 'furaptra') (9) is in fact the most popular indicator for monitoring $[Ca^{2+}]$ changes in the internal stores. It generally loads very well into cells and is quite bright, an important consideration for measurements of fluorescence from small subcellular structures. Mg^{2+} sensitivity is an oft-cited hazard

Table 1 Low-affinity fluorescent indicators that have been used to measure [Ca^{2+}] in organelles[a]

Indicator	K_d for Ca^{2+} (μM)	Excitation (nm) Emission (nm)	Comments
Mag-fura-2 ('furaptra')	53	345/375 ex. 510 em.	Ratiometric
Mag-fura-5	28	340/380 ex. 510 em.	Ratiometric
Fura-2 FF	35	340/380 ex. 510 em.	Ratiometric; available from TefLabs
Mag-Indo-1	32	351 ex. 405/485 em.	Ratiometric (emission)
Mag-fura red	17	488 ex. 630 em.	Fluorescence decreases upon Ca^{2+} binding
Fluo-3 FF	41	515 ex. 530 em.	Available from TefLabs; fluorescence increases upon Ca^{2+} binding
Oregon Green BAPTA-5N	20	492 ex. 521 em.	Excited efficiently by 488 nm laser line; fluorescence increases upon Ca^{2+} binding

[a] Dyes are available from Molecular Probes Inc. unless otherwise noted.

of using this probe, although in practical terms, this has presented minor problems (16, 21). Dyes that are insensitive to Mg^{2+} are also available (e.g. fura-2 FF).

2 Experimental protocols (permeabilized cells)

Detailed descriptions of protocols for selective permeabilization of the plasma membrane using digitonin, streptolysin O, and alpha toxin, as well as methods for conducting experiments in permeabilized cells have appeared elsewhere (22), and will be summarized only briefly here.

Protocol 1

Measurements of calcium in internal stores of digitonin permeabilized cells

Reagents

- 5–10 mM stock solution of dye-AM in anhydrous DMSO
- Digitonin stock solution: 5 mg/ml in H_2O
- 'Intracellular buffer' (in mM): 125 KCl, 25 NaCl, 10 Hepes, 1 Na_2ATP, 0.2 $MgCl_2$, 200 μM $CaCl_2$, 500 μM EGTA to give a final free [Ca^{2+}] of approx. 100 nM, pH 7.25
- KCl rinse solution (in mM):125 KCl, 25 NaCl, 10 Hepes 0.2 $MgCl_2$ pH 7.25
- Calibration solution: supplement KCl rinse solution with Ca^{2+} ionophores ionomycin or 4Br-A23187 (10 μM final concentration)

Protocol 1 continued

Method

As mentioned above, different cell types have distinctive dye handling characteristics. Optimizing compartmentalization requires some experimentation with loading conditions, including dye concentration, loading and rinsing time, and temperature. Exposure to 2 μM of the AM ester form of the dye for 30 min at 37 °C has yielded good results in many cell types. Longer incubation times do not always produce better dye trapping.

1. Load cells with the AM form of the chosen fluorescent indicator. Add the appropriate volume of indicator-AM stock solution directly to cells that are either in suspension or grown on coverslips and mix gently. Cultured cells can be put back in the tissue culture incubator for loading at 37 °C.
2. Briefly wash cells with KCl rinse solution.
3. Replace rinse solution with intracellular buffer containing 5 μg/ml digitonin and leave for the appropriate amount of time (usually 1–2 min at 37 °C). Cells should begin to lose the dye from the cytoplasm, generally resulting in a 50–90% drop in fluorescence intensity as measured at the Ca^{2+}-insensitive wavelength.
4. Switch immediately to the digitonin-free intracellular buffer to avoid damage to organelle membranes. The nucleus (which does not contain organelles) will be readily apparent as a non-fluorescent central region while dye remaining in the periphery should have a 'reticular' distribution. Procedures for fluorescence measurements of organelles-trapped indicator are essentially the same as for dyes trapped in the cytoplasm (see Chapter 2).

It is noteworthy to mention here that many investigators use digitonin-permeabilization of the plasma membrane in order to assess the degree of dye compartmentalization of cytosolic Ca^{2+} probes such as fura-2. These controls for compartmentation artefacts are sometimes conducted using rather generous concentrations of detergent (up to 100 μg/ml), and are carried out in normal NaCl Ringer's solution. We have found that such harsh permeabilization conditions greatly diminish the retention of fluorophores in organelles (Hofer and Machen, unpublished observations). Care must be taken to minimize both the time and digitonin concentration as much as possible for both measurements of internal store $[Ca^{2+}]$ and for reliable estimates of organelle-bound dye.

2.1 Notes on calcium buffers

A computer program that allows the calculation of free $[Ca^{2+}]$ in a solution containing multiple ligands is indispensable (refer also to the informative chapter by Bers and Patton; ref. 23). A particularly good program is MaxChelator, which can now be accessed through the internet: (http://www.stanford.edu/~cpatton/maxc.html).

The choice of Ca^{2+} buffer (i.e. BAPTA versus EGTA versus unbuffered) in intracellular-like solutions may be an important consideration in experiments designed to examine the kinetics of $InsP_3$-induced Ca^{2+} release (24). For example, oscillations in luminal ER $[Ca^{2+}]$ were recently observed by Hajnóczky and Thomas (25) and independently by Tanimura and Turner (26) using permeabilized cell systems. The key to reproducing these fluctuations experimentally was the use of specially prepared unbuffered intracellular medium that permitted local Ca^{2+} feedback to occur at the level of the $InsP_3$ receptor. Since the contaminating $[Ca^{2+}]$ in a nominally Ca^{2+}-free solution can range from 1–10 μM, these solutions require special preparation; these procedures have been described previously (27).

2.2 Experiments in patch clamped cells

An alternative strategy for ridding dye from the cytoplasm dye takes advantage of the whole-cell patch clamp technique to dialyse cytosolic indicator. Simultaneous measurements of I_{CRAC} (a Ca^{2+} current activated by the depletion of internal stores) and the store $[Ca^{2+}]$ have been made concurrently in mag-fura-2-loaded RBL-1 cells using this method (28). Others have used this general approach to examine the precise temporal relationship between ER $[Ca^{2+}]$ dynamics and Ca^{2+} signalling events in the cytoplasm. Cytoplasmic $[Ca^{2+}]$ changes have been assayed either indirectly as the activation of Ca^{2+}-sensitive Cl^- or K^+ currents (4, 12), or directly by introducing impermeant high affinity Ca^{2+} dyes such as fluo-3 free acid via the patch pipette (29). One difficulty that may be encountered when combining this approach with electrophysiological measurements is run down of the whole-cell current, a problem that is exacerbated by the fact that protracted dialysis times (several minutes) may be required to eliminate dye entirely from the cytoplasm.

2.3 Intact cell measurements

It is well known that cytoplasmically-trapped indicators are eventually lost from the cytosol in many cell types. This export may occur through the action of currently unidentified membrane transporters susceptible to organic anion transport inhibitors such as sulfinpyrazone and probenecid (30, 31). In some cell types, dye is evidently extruded from the cytoplasmic compartment more rapidly than from internal stores (13–15). Thus there exists a 'window' of time where selective loading of organelles is achieved, allowing non-invasive measurements of store $[Ca^{2+}]$ in single intact cells. Although this phenomenon proceeds efficiently at 37 °C, it is further enhanced by increasing the incubation temperature to 40 °C.

How do Ca^{2+} release kinetics in intact and permeabilized cells differ? In the experiment depicted in *Figure 1*, BHK-21 fibroblasts were loaded with 2 μM mag-fura-2 AM for 50 minutes in a CO_2 tissue culture incubator at 37 °C. The appearance of the cells strongly suggested selective retention of dye within organelles (see ref. 14). Moreover, the resting mag-fura-2 ratio was relatively high, indicating that most of the dye was in an environment of elevated $[Ca^{2+}]$.

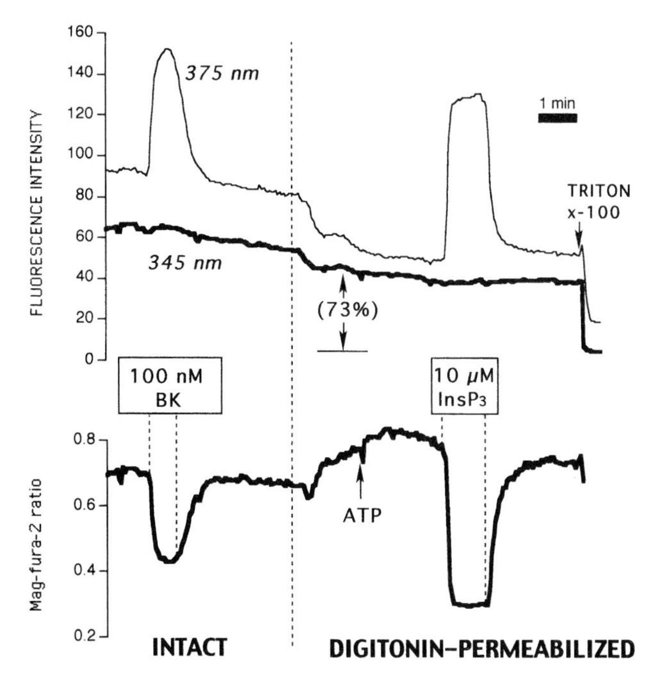

Figure 1 Fluorescence measurement of mag-fura-2 loaded BHK-21 cell. Comparison of responses before and after permeabilization with digitonin. See text for details.

A direct comparison of the response to a native Ca^{2+}-mobilizing agonist (bradykinin; 100 nM) in the intact cell, and to a supramaximal dose of $InsP_3$ (10 μM) following plasma membrane permeabilization of the same cell is shown (*Figure 1*; adapted from a record published in ref. 14). The upper panel represents the individual fluorescence intensities at 345 nm (Ca^{2+}-insensitive) and 375 nm (the wavelength at which $[Ca^{2+}]$ is inversely proportional to fluorescence intensity). In this particular cell, digitonin treatment released 27% of the intracellular probe (as measured by the decrease at 345 nm isoexcitation wavelength). In most cells, the digitonin-sensitive drop in fluorescence was even less (17% on average, with many cells releasing no dye whatsoever), indicating that the large majority of dye was contained within digitonin-resistant subcellular compartments. The kinetic characteristics and magnitude of the response to Ca^{2+} mobilizing agents in the intact and permeabilized states displayed subtle differences. The discrepancy in magnitude may be accounted for by the mag-fura-2 retained within the cytoplasm of the intact cell, which can disproportionately suppress the measured response for reasons outlined in ref. 16. Alternatively, the physiological agonist may generate less $InsP_3$ than was applied exogenously to the permeabilized cells in the second part of the experiment. Myriad other factors may account for the kinetic variation in the two responses, e.g. buffering at the cytosolic face, rapid bradykinin receptor desensitization, or metabolism of second messengers.

The possibility of following changes in free $[Ca^{2+}]$ in the ER of single intact cells following physiological stimulation with native agonists offers obvious advantages. Unfortunately, attempts to reproduce these same types of measurements have not always proven successful for all cell types (among these, intact gastric epithelial, HEK 293, RBL-1, and NG-108 cells). The molecular identification of the putative transporters involved in the dye redistribution with prolonged loading could, in principle, be used to render a wide range of transfectable cell types amenable to this approach.

2.4 Calibration procedures

Approaches for converting dye fluorescence into actual free [Ca^{2+}] have been described previously in detail (16, 22), and will only be covered superficially here. The methods are quite similar to those used for calibrating the high-affinity counterparts of these dyes (Chapter 2, this volume). Methods include:

(a) *In vitro* calibrations. A sample of the free acid form of the dye is placed on the microscope stage and the fluorescence is measured in the presence of zero and saturating [Ca^{2+}]. These values are correlated to those observed in the living cell. This approach is generally not as reliable as the following two approaches, due to differences in the K_d of the indicator *in situ* versus *in vitro*.

(b) R_{min} / R_{max} method for ratiometric dyes (or the corresponding technique for single wavelength indicators; see Chapter 2, this volume).

(c) *In situ* calibration. Because relatively high [Ca^{2+}] are involved, the fluorescence response of the dye can be monitored when free [Ca^{2+}] has been clamped using ionophores and external solutions of known [Ca^{2+}]. This is the most accurate method, as it is independent of knowledge of the K_d of the fluorophore *in situ*, but assumes that ionophores can effectively equilibrate internal and external [Ca^{2+}].

Note that the calibration cannot provide an absolute measure of [Ca^{2+}] unless dye is contained only in the organelle of interest (16). It is useful however, in that it can correct for the non-linearity of the dye response.

3 Pitfalls

All of the same perils faced by investigators using fluorescent indicators to measure cytoplasmic [Ca^{2+}] (e.g. artefacts due to viscosity, pH, heavy metals, photodamage, and photodegradation of dyes) (30) can also plague the use of compartmentalized indicators. The main disadvantage is the indiscriminate nature of dye trapping, and in this regard genetically encoded probes (Chapters 1 and 4) are far superior. Other problems specific to the trapped dye technique are discussed below.

3.1 Problems of multiple compartments

BHK-21 cells have proven to be a useful model for studying Ca^{2+} handling of internal stores using compartmentalized fluorescent indicators. But even when using a cell type in which the majority of the probe is contained in the agonist-sensitive store, it is essential to keep in mind the potential contributions of dye in other organelles. When using a new cell system, it is vital to define the compartments in which the dye measures [Ca^{2+}] changes for proper interpretation of the data. As detailed previously by Hofer and Machen (11) and Hofer and Schulz (16), the presence of multiple, independent, dye-containing compartments introduces a particular non-linearity in the fluorescence measurement. This may limit the quantitative accuracy of the method for certain applications, but important qualitative information can still be obtained.

3.2 Interference by Mg^{2+} and other ions

Many of the more popular low-affinity Ca^{2+} indicators (mag-Indo-1, mag-fura-2) were, as their names would suggest, originally designed as fluorescent probes for Mg^{2+}. The Mg^{2+} sensitivity of these dyes must be taken into account for measurements of internal store [Ca^{2+}]. In several cell systems Mg^{2+} interference was shown to be minimal (11, 16, 21). Of much greater concern is potential interference by heavy metals such as Zn^{2+}, which have incredibly high affinity for many of the popular indicators (32, and Chapter 3).

4 Advantages

The compartmentalized dye approach offers several advantages over other optical techniques for monitoring $[Ca^{2+}]_{ER}$, such as those incorporating calmodulin and green fluorescent proteins ('chameleons'; see Chapter 1) or targeted recombinant photoproteins (ER-targeted aequorins; see Chapter 4). Unlike these two methods that are based on genetically encoded probes, the trapped low affinity indicators can be used in cell preparations not amenable to transfection, such as acutely isolated cells or tissues. The second main advantage is that the technique is readily adaptable to existing instrumentation already available in many laboratories using the high-affinity counterparts of the dyes used for this technique (i.e. fura-2, fluo-3, Indo-1). With judicious selection of cell types and fluorescent probes it is possible to make sensitive and selective measurements of [Ca^{2+}] changes in internal stores.

5 Manipulating intrastore [Ca^{2+}] with a membrane-permeant Ca^{2+} chelator, TPEN

A standard tool long used by physiologists studying intracellular pH regulation has been the 'ammonium prepulse' technique (33). Ammonia (NH_3) is a weak base that readily penetrates cell membranes, while the protonated species (NH_4^+) does not. Cells exposed to NH_3/NH_4^+ will initially alkalinize. Continued exposure to the weak base allows protons (originating from endogenous acid loading processes) to accumulate in the cell, because they are rapidly picked up by the NH_3 to form NH_4^+. When ammonia/ammonium are washed from the

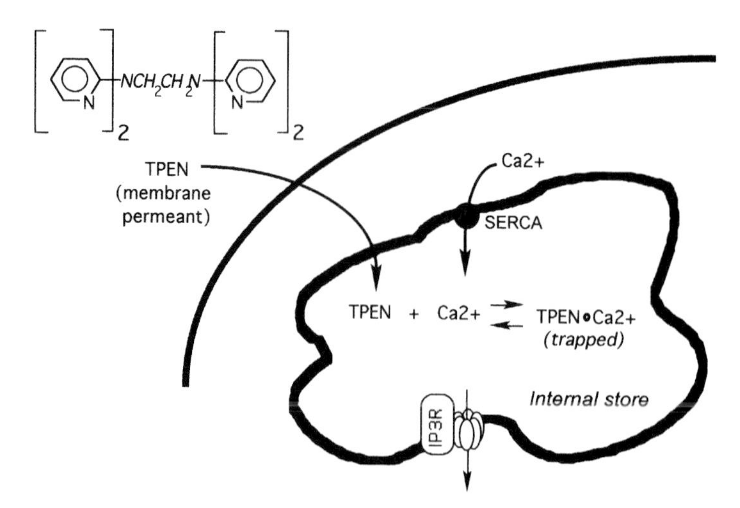

Figure 2 TPEN is membrane-permeant in its free form, and becomes trapped within stores upon complexation with Ca^{2+}.

external bath, there is a profound acidification of the cell interior as NH_3 rapidly exits, driving the dissociation of NH_4^+ and H^+, leaving a 'load' of protons trapped within the cell.

TPEN, a membrane-permeant compound originally described for use as a scavenger of heavy metals (34), can be used in an analogous way to control the free $[Ca^{2+}]$ of internal stores. The chelator has an affinity for Ca^{2+} that lies in a range suitable for manipulating the intraluminal $[Ca^{2+}]$ (K_d between 40 and 130 μM). Like ammonia, TPEN is membrane-permeant in its neutral, uncomplexed form, but becomes charged when it binds Ca^{2+}. Thus when free TPEN partitions into compartments of high $[Ca^{2+}]$ such as the internal store, it chelates Ca^{2+}, lowering the free $[Ca^{2+}]$ and becoming trapped in the process (see *Figure 2*).

As shown in *Figure 3*, this compound can be used to reversibly lower the free intraluminal $[Ca^{2+}]$ in a dose-dependent fashion, as measured by the mag-fura-2 ratio of permeabilized BHK-21 fibroblasts. Shown for comparison is the response to a supramaximal dose of $InsP_3$.

The effects of TPEN on the free $[Ca^{2+}]$ in internal stores of *intact* BHK-21 fibroblasts are illustrated in *Figure 4*. In the first part of the record, mag-fura-2-loaded cells were exposed to a solution containing 2 mM total TPEN and 2 mM total Ca^{2+}. The free $[Ca^{2+}]$ and free [TPEN] in this solution were calculated to be 435 μM each, with the remainder existing as the 'inactive' (i.e. membrane-impermeant) TPEN:Ca^{2+} complex. This treatment resulted initially in a relatively small, rapid drop in the mag-fura-2 ratio as free TPEN entered the store and became complexed with Ca^{2+}. In the continued presence of the chelator, there was a slow recharging of the store $[Ca^{2+}]$, presumably due to compensatory pumping by resident SERCAs. Noteworthy is the fact that when extracellular TPEN was washed from the cells, there was an overshoot in intraluminal $[Ca^{2+}]$, followed by an eventual recovery back to the baseline value. Thus TPEN can be used not

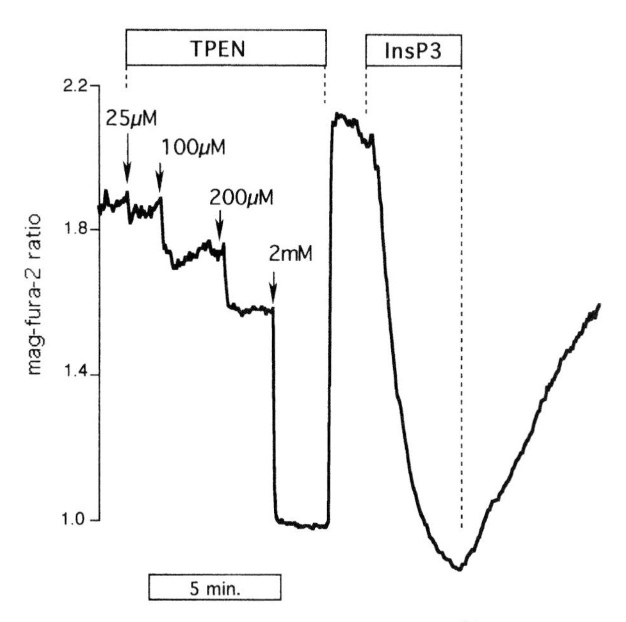

Figure 3 TPEN dose-dependently lowers free Ca^{2+} in internal stores of digitonin-permeabilized BHK-21 cells.

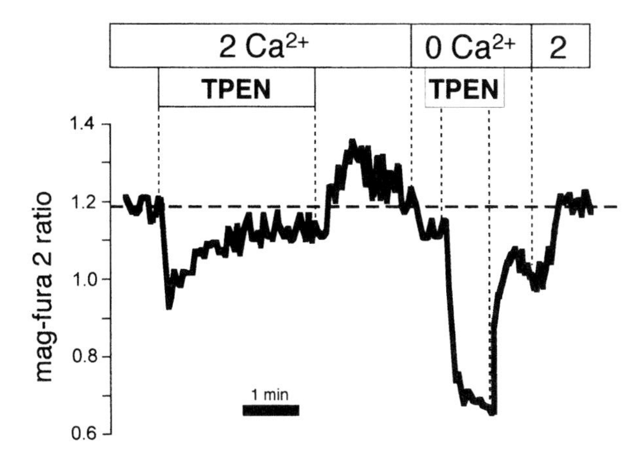

Figure 4 Mag-fura-2 ratio from intact BHK-21 cell. Application of TPEN in the presence of Ca^{2+} in the external bath results initially in a lowering of intraluminal $[Ca^{2+}]$ and an overshoot when the chelator is washed off. See text for details.

only to reduce the free intraluminal $[Ca^{2+}]$ in a dose-dependent manner, but also to 'hyperload' internal stores, provided there is adequate extracellular Ca^{2+} available for refilling. In the same experiment 2 mM TPEN was applied in a nominally Ca^{2+}-free Ringer's solution. As expected, this resulted in a much larger drop in the free $[Ca^{2+}]_{ER}$, since the free ('active') [TPEN] was much greater under these conditions.

5.1 Practical considerations for using TPEN

A stock solution of TPEN with a maximal concentration of around 200 mM can be prepared by dissolving the appropriate amount of the compound in 100% EtOH. It is also possible to dissolve TPEN crystals directly in experimental solutions. This is important in cases where relatively high final concentrations (greater than 1 mM) of the chelator are desired, so as to avoid the deleterious effects of EtOH in the buffer. Be advised, however, that TPEN goes into aqueous solution very slowly, so it is best to prepare this buffer one day in advance.

As mentioned above, most Ca^{2+} indicators are quite sensitive to heavy metals. It is important to distinguish between fluorescence changes caused by scavenging of endogenous heavy metals versus those resulting from Ca^{2+} chelation when using TPEN. Given the extraordinary affinity of the compound for heavy metals (e.g. K_d for $Zn^{2+} = 2.6 \times 10^{-16}$ M), low doses (10 μM) of TPEN are sufficient to assess the contribution of Zn^{2+}, Cd^{2+}, Cu^{2+} etc. to the overall signal without appreciably influencing $[Ca^{2+}]$ (32, 34, see also Chapter 3 of this volume). For mag-fura-2, the ratio may or may not decrease following application of the chelator, while the individual fluorescence intensities will increase at both wavelengths as the TPEN competes away heavy metals bound to the dye.

5.2 Applications

Other strategies for manipulating luminal $[Ca^{2+}]$, such as the use of ionophores, SERCA inhibitors, or agents that act on Ca^{2+} release channels (InsP3, cADP ribose, sphingosine, etc.) invariably result in release of Ca^{2+} into the cytoplasm. It can be difficult to obtain graded responses and many of these agents are irreversible. Moreover, none of these treatments can be used to hyperload the store with Ca^{2+}. TPEN, on the other hand, is a versatile tool for rapidly and reversibly clamping the free $[Ca^{2+}]_{ER}$ independent of Ca^{2+} release. The compound has been used recently to lower the free $[Ca^{2+}]$ in internal stores to reversibly activate store-operated Ca^{2+} currents in fibroblasts and RBL-1 cells (28) and in *Xenopus* oocytes (35). The quantitative relationship between store depletion and the activation of capacitative Ca^{2+} entry was investigated (28). These studies provided confirmation that the lowering of free $[Ca^{2+}]$ *per se* is sufficient to activate Ca^{2+} entry. It also allowed a unique opportunity to study the kinetics of current inactivation. Ordinarily, once the Ca^{2+}-mobilizing agonist is removed and the stimulus terminated, several minutes elapse before stores are fully recharged with Ca^{2+}, causing Ca^{2+} entry to cease. Under these conditions, it is possible that the refilling of internal stores is the rate-limiting step in the inactivation of I_{CRAC}. The intrinsic inactivation of the current was in fact shown to occur with a faster time course than the refilling of stores, but more slowly than the activation of the current.

Many other biological applications for low affinity membrane-permeant Ca^{2+} chelators can be envisaged. For example, TPEN can potentially be used to probe the intrinsic Ca^{2+} buffering capacity of the internal store, or to examine how luminal $[Ca^{2+}]_{ER}$ affects Ca^{2+} leaks and the kinetics of $InsP_3$-induced release. In addition, there is growing evidence (3) that intraluminal ER Ca^{2+} may participate in regulating a number of important cell functions (e.g. protein folding and sorting, stress responses, and cell growth). Methods that allow selective manipulation of [Ca^{2+}] within the ER should prove useful for understanding how free calcium ions participate in these key biological processes.

Acknowledgements

The author is grateful to Tullio Pozzan for stimulating discussions, and to Dr David I. Soybel for providing the resources used for preparing this manuscript. A. M. H. was supported by the Brigham Surgical Group Foundation and the Medical Research Service of the Veterans Administration. Portions of this work were funded by a grant to A. M. H. from the Harvard Digestive Diseases Center.

References

1. Grynkiewicz, G., Poenie, M., and Tsien, R. Y. (1985). *J. Biol. Chem.*, **260**, 3440.
2. Tsien, R. Y. (1989). *Methods Cell Biol.*, **30**, 127.
3. Pozzan, T., Rizzuto, R., Volpe, P., and Meldolesi, J. (1994). *Physiol. Rev.*, **74**(3), 595.
4. Mogami, H., Tepikin, A. V., and Petersen, O. H. (1998). *EMBO J.*, **17**(2), 435.
5. Terasaki, M. and Sardet, C. (1991). *J. Cell Biol.*, **115**, 1031.
6. Connor, J. A. (1993). *Cell Calcium*, **14**, 185.
7. Glennon, M. C., Bird, G. St. J., Takemura, H., Thastrup, O., Leslie, B. A., and Putney, J. W. Jr. (1992). *J. Biol. Chem.*, **267**(35), 25568.
8. Short, A. D., Klein, M. G., Schneider, M. F., and Gill, D. L. (1993). *J. Biol. Chem.*, **268**(34), 25887.
9. Raju, B., Murphy, E., Levy, L. A., Hall, R. D., and London, R. E. (1989). *Am. J. Physiol.*, **256**, C540.
10. Hofer, A. M. and Machen, T. E. (1993). *Proc. Natl. Acad. Sci. USA*, **90**, 2598.
11. Hofer, A. M. and Machen, T. E. (1994). *Am. J. Physiol.*, **267**, G442.
12. Tse, F. W., Tse, A., and Hille, B. (1994). *Proc. Natl. Acad. Sci. USA*, **91**, 9750.
13. Landolfi, B., Curci, S., Debellis, L., Pozzan, T., and Hofer, A. M. (1998). *J. Cell Biol.*, **142**(5), 1235.
14. Hofer, A. M., Landolfi, B., Debellis, L., Pozzan, T., and Curci, S. (1998). *EMBO J.*, **17**(7), 1986.
15. Golovina, V. A. and Blaustein, M. P. (1997). *Science*, **275**(5306), 1643.
16. Hofer, A. M. and Schulz, I. (1996). *Cell Calcium*, **20**(3), 235.
17. Gerasimenko, O. V., Gerasimenko, J. V., Belan, P. V., and Petersen, O. H. (1996). *Cell*, **84**(3), 473.
18. Satoh, T. and Hosokawa, M. (1998). *Annu. Rev. Pharmacol. Toxicol.*, **38**, 257.
19. Nelson, E. J., Zinkin, N. T., and Hinkle, P. M. (1998). *Cancer Chemother. Pharmacol.*, **42**(4), 292.
20. Hofer, A. M., Schlue, W. R., Curci, S., and Machen, T. E. (1995). *FASEB J.*, **9**, 788.
21. Sugiyama, T. and Goldman, W. F. (1995). *Am. J. Physiol.*, **269**, C698.
22. Hofer, A. M. (1998). In *Methods in molecular biology: calcium signaling protocols* (ed. D. G. Lambert), p. 249. Humana Press, New Jersey, USA.
23. Bers, D. M. and Patton, C. W. (1994). *A practical guide to the study of calcium in living cells* (ed. R. Nuccitelli). In *Methods in cell biology*, Vol. 40. Academic Press, San Diego.

24. Montero, M., Barrero, M. J., and Alvarez, J. (1997). *FASEB J.*, **11**(11), 881.
25. Hajnóczky, G. and Thomas, A. P. (1997). *EMBO J.*, **16**(12), 3533.
26. Tanimura, A. and Turner, R. J. (1996). *J. Biol. Chem.*, **271**, 1996.
27. Morgan, A. J. and Thomas, A. P. (1998). In *Methods in molecular biology: calcium signaling protocols* (ed. D. G. Lambert), p. 93. Humana Press, New Jersey, USA.
28. Hofer, A. M., Fasolato, C., and Pozzan, T. (1998). *J. Cell Biol.*, **140**(2), 325.
29. Chatton, J.-Y., Liu, H., and Stucki, J. W. (1995). *FEBS Lett.*, **368**, 165.
30. Roe, M. W., Lemasters, J. J., and Herman, B. (1990). *Cell Calcium*, **11**, 63.
31. Di Virgilio, F., Steinberg, T. H., and Silverstein, S. C. (1990). *Cell Calcium*, **11**, 57.
32. Snitsarev, V. A., McNulty, T. J., and Taylor, C. W. (1996). *Biophys. J.*, **71**, 1048.
33. Boron, W. F., Roos, A., and De Weer, P. (1978). *Nature*, **274**(5667), 190.
34. Arslan, P., Di Virgilio, F., Betrame, M., Tsien, R. Y., and Pozzan, T. (1985). *J. Biol. Chem.*, **260**, 2719.
35. Yao, Y., Ferrer-Montiel, A. V., Montal, M., and Tsien, R. Y. (1999). *Cell*, **98**(4), 475.

Chapter 7

Measuring calcium in the nuclear envelope and nucleoplasm

Oleg Gerasimenko and Julia Gerasimenko
The Physiological Laboratory, The University of Liverpool, Crown Street, Liverpool L69 3BX, UK.

1 Introduction

1.1 Techniques and experiments described in the chapter

The first part of this chapter describes measurements of free calcium concentration in isolated nuclei using fluorescent probes. It was established that the Ca^{2+}-sensitive fluorescent probe fura-2 is concentrated in the nuclear envelope and that distribution of fura-2 fluorescence is similar to that of the endoplasmic reticulum marker (1–3). This makes it possible to measure Ca^{2+} concentration in the nuclear envelope. The dextrans labelled with fluorescent Ca^{2+} indicators (Calcium Green-1 and fura-2) were distributed uniformly in the nucleoplasm and can be used to monitor changes of Ca^{2+} concentration in the nucleoplasm (1).

The second part of this chapter describes the method that has been used to compare Ca^{2+} transients in the nucleus and other regions of cytosol in isolated cells (4).

1.2 Calcium transport in the nucleus

Cytosolic Ca^{2+} spiking is induced by a variety of extracellular agents (neurotransmitters, hormones, or metabolites) and has arisen as an important signalling mechanism involved in regulation of cellular events including secretion, contraction, motility, and metabolism (5, 6). Considering the importance of local calcium signalling (7) it seems desirable to investigate the Ca^{2+} transport pathways that exist in the nucleus.

It is generally established that the outer nuclear membrane has endoplasmic reticulum (ER) properties and is continuous with the ER membranes. The protein composition of the inner nuclear membrane is different from that of the outer nuclear membrane (8). The space between the outer and inner nuclear membrane is an extension of the ER lumen.

The nuclear pore complexes are large supramolecular assemblies (approximately 125 MDa) which may contain more than 100 different protein components called nucleoporins. These complexes mediate the traffic between the nucleoplasm and the cytoplasm (9, 10). The nuclear pore complex is generally accepted to be very permeable to substances with a molecular weight of up to 10–20 kDa (11), whereas most proteins and ribonucleic acids are too large to diffuse through the nuclear pore complex at physiologically relevant rates. Such large molecules are transported by a temperature and energy-dependent process, that requires the activity of specific nuclear pore complex proteins as well as cytosolic transport (10).

Intracellular Ca^{2+} changes are important for the control of fundamental nuclear processes such as gene expression, DNA replication and repair, phosphorylation of nuclear proteins, and apoptosis (12). Nuclei can accumulate Ca^{2+} via the endoplasmic reticulum type ATP-dependent Ca^{2+} pump (13). Nuclei contain the Ca^{2+} binding protein calmodulin and the main binding protein of the endoplasmic reticulum, calreticulin (14, 15). It was found that nuclei possess the functional receptors for the Ca^{2+}-releasing messenger inositol 1,4,5-trisphosphate (16–18), and the mechanism for production of this messenger (19).

In stimulated cells some differences of Ca^{2+} concentration in the nucleoplasm and the cytosol have been observed (5). Other studies suggested that amplification of changes in free Ca^{2+} concentration in the nucleus, is a measurement artefact and differences between cytosolic and nucleoplasmic Ca^{2+} concentrations equilibrate rapidly (20–22).

Due to the high permeability of the nuclear pore complex to Ca^{2+} and the many possible actions of free calcium inside the nucleus, it would seem desirable that cytosolic calcium signals designed to stimulate actions at the cell membrane (for example secretion by exocytosis) should not be transmitted to the nucleoplasm. For example there are local Ca^{2+} signals in the secretory pole of pancreatic acinar cells that are sufficient to stimulate exocytosis but do not spread to the nuclear area (23, 24).

The dynamic changes of free calcium concentration in the cytosol and nucleus can be investigated using confocal microscopy and fluorescent Ca^{2+}-sensitive indicators.

2 Experimenting with isolated nuclei

Section 2.1 will describe the isolation of nuclei. The description of loading of nucleoplasm and nuclear envelope with calcium indicators is given in Section 2.2. Section 2.2 also considers calcium measurements in isolated nuclei.

2.1 Preparation of isolated mouse liver nuclei

Single nuclei can be isolated from mouse liver using the method of Nicotera *et al.* (25). We used a slightly modified procedure; *Protocol 1* describes the technique that we used for isolation of nuclei. The main modification was that the last high-speed centrifugation was omitted. In our experience the modified procedure yields less leaky nuclei. The final pellet of isolated nuclei was resuspended in standard buffer and loaded with fluorescent dye, then observed using confocal microscopy (see *Protocols 1–3*).

Protocol 1

Preparation of single isolated mouse liver nuclei

Equipment and reagents

- Homogenizer (glass/Teflon) (Sigma)
- Cheesecloth
- Centrifuge (capable of 1000 g)
- Isolated mouse liver
- Standard solution: 125 mM KCl, 2 mM K_2HPO_4, 50 mM Hepes, 0.1 mM EGTA, 1 mM ATP, 1 mg/ml trypsin inhibitor (from soybean type I-S), 0.25 mM sucrose pH 7.0 (all chemicals from Sigma)

Method

1. Wash the liver isolated from a male CD1 mouse in the standard solution (supplemented with 0.25 mM sucrose and 1 mg/ml trypsin inhibitor, see above). Cut into small pieces and homogenize with glass/Teflon homogenizer (eight to ten strokes) in the same buffer.[a]
2. Filter the homogenate through several layers of cheesecloth.
3. Centrifuge the homogenate at 1000 g for 1 min and resuspend the nuclear pellet in the standard solution with 0.25 mM sucrose and 1 mg/ml trypsin inhibitor.
4. Homogenize the suspension of the nuclei with glass/Teflon homogenizer (five strokes).
5. Centrifuge the homogenate at 1000 g for 1 min and gently resuspend the nuclear pellet in the standard buffer.

[a] All steps at 4 °C.

2.2 Fluorescent labelling of nucleoplasm and nuclear envelope of isolated nuclei: measuring calcium signals in the nuclear envelope and nucleoplasm

Plate 7a illustrates the ring-like distribution of the fluorescence of fura-2 (loaded in AM form) in isolated nuclei. This fluorescent indicator is localized in the areas of the nuclear envelope and ER (which remains connected to the nuclei after the isolation procedure). In this experiment the excitation wavelength was 363 nm (argon laser line close to the isobestic point of fura-2). The distribution of the fluorescence in the nucleus obtained using the ER marker $DiOC_6(3)$ is similar to that of fura-2 (please see Plate 7b).

The ability of dextrans bound to calcium indicators to accumulate in the nucleoplasm could be used to measure Ca^{2+} concentration in this region of the cell. The distribution of fluorescence intensity of the DNA–RNA marker ethidium bromide (see Plate 7d) is similar to the distribution of the fluorescence of the dextrans labelled with fluorescent Ca^{2+} indicators such as fura-2 dextran (see Plate 7c) and Calcium Green-1 dextran.

Protocol 2

Labelling of isolated nuclei with fura-2 AM

Equipment and reagents

- Confocal microscope and/or non-confocal imaging system
- Isolated mouse liver nuclei in standard buffer (see *Protocol 1*)
- Fura-2 AM, $DiOC_6(3)$ (both from Molecular Probes)

Method

1. Incubate the nuclei obtained from one liver with standard buffer (see *Protocol 1*) with 20 μM of fura-2 AM added at 4 °C for 45 min. Wash loaded nuclei by centrifugation at 1000 *g* for 1 min and resuspend in 1–2 ml of standard buffer.
2. Place samples of nuclei loaded with fluorescent indicator in the perfusion experimental chamber at room temperature and wash them by perfusion with standard solution for a few minutes before the beginning of the experiment.
3. Use the wavelength of excitation close to isobestic point for fura-2 (e.g. 363 nm argon laser line) to observe accumulation of fura-2 in the nuclear envelope. Use 340 and 380 nm wavelengths of excitation to resolve calcium responses in the nuclear envelope; emission can be recorded using a 505 filter.
4. Stain nuclei with $DiOC_6$ to confirm the localization of the probe to the nuclear envelope. We used $DiOC_6$ at a concentration of 2.5 μg/ml. Staining was performed by incubation for 5 min at room temperature. After the incubation nuclei were washed once by centrifugation at 1000 g for 1 min with standard buffer, resuspended in standard buffer, and placed in the experimental chamber of the confocal system. $DiOC_6$ fluorescence was excited at 488 nm and emission was collected at $>$ 520 nm.

In our experiments with fura-2 AM loaded nuclei, the fluorescence of fura-2 in an isolated single nucleus did not change when the external Ca^{2+} concentration increased from 10^{-8} M to 1 mM. However, addition of 20 μM of ionomycin to an external solution with a high Ca^{2+} concentration (1 mM) resulted in saturation of the fura-2 ratio signal. A subsequent wash with the Ca^{2+} chelator, EGTA, in the presence of ionomycin induced recovery of the signal followed by a continuous slow decline of the nuclear Ca^{2+} concentration (*Figure 1A*). *Figure 1B* represents an ATP-dependent accumulation of Ca^{2+} into the nuclei.

Accumulated calcium in the nuclei could be released by IP_3, but this effect was difficult to detect in the ATP-containing solution. After removal of ATP from

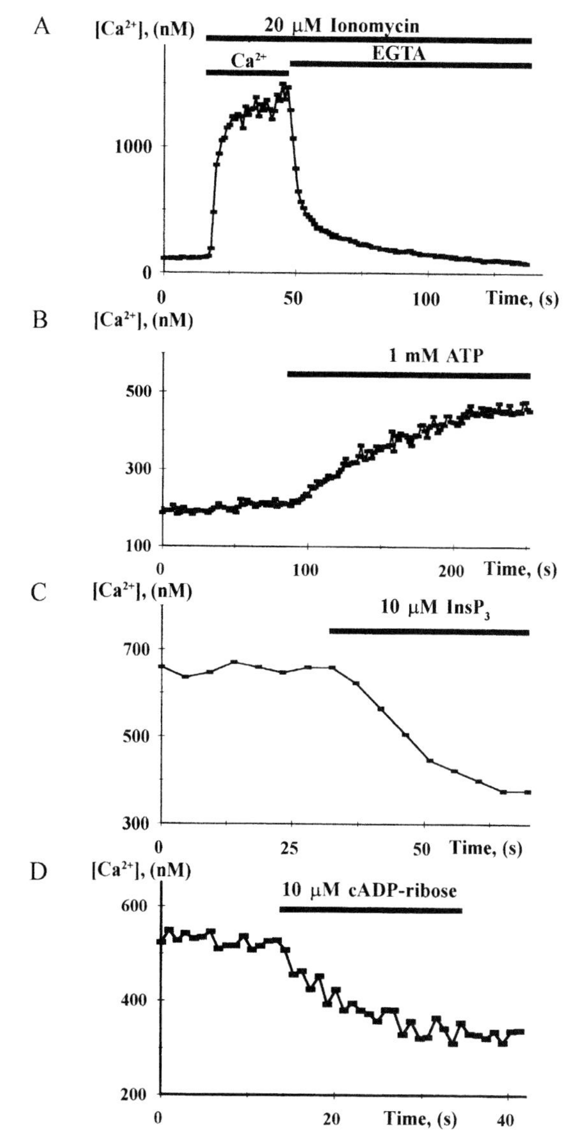

Figure 1 Free calcium measurements in nuclear envelope (nuclei loaded with fura-2 AM). (A) The calibration of fura-2 loaded nucleus with high Ca^{2+} (1 mM) and EGTA (2 mM) in the presence of 20 μM of ionomycin. (B) ATP-dependent calcium uptake into an isolated nucleus. External standard solution contained 100 nM of free calcium. (C) Application of IP_3 induced reduction of free calcium in the nucleus. (D) Application of cADP-ribose induced reduction of free calcium in the nucleus. Reproduced with permission from Cell Press (1).

the external solution, IP_3 induced a significant decrease in the nuclear Ca^{2+} concentration (*Figure 1C*). Similar experiments have been carried out by Nicotera *et al.* (16, 25) and Hechtenberg *et al.* (26) in suspensions of isolated nuclei and showed the existence of ATP-dependent accumulation of Ca^{2+} as well as IP_3- and calmidazolium-induced Ca^{2+} release. In our experiments we have controlled the distribution of the Ca^{2+} indicator, and we have found that all these effects occur in isolated nuclei, with a rim distribution of fura-2. Therefore Ca^{2+} accumulation and release occur into and from the stores around the nuclear envelope.

The role of cADP-ribose as a calcium-releasing messenger has been previously reported in different systems (27–30). *Figure 1D* illustrates that cADP-ribose (10 μM) induced loss of calcium from the nuclear envelope.

Protocol 3

Labelling the nucleoplasm with dextrans of calcium indicators; recording the nucleoplasmic calcium signals

Equipment and reagents

- Fura-2 dextran, Calcium Green-1 dextran, ethidium bromide (all reagents from Molecular Probes)
- Confocal microscope and/or non-confocal imaging system

Method

1. Incubate the nuclei with 20 μM of fura-2 dextran or Calcium Green-1 dextran for 10 min at 4 °C in the standard bath solution. Wash the nuclei by centrifugation at 1000 g for 1 min to reduce the background fluorescence and then resuspend in standard buffer. Fluorescent dextrans accumulate in the nuclei (please see Plate 7c) and remain there for tens of minutes.
2. Place samples of nuclei loaded with fluorescent indicator in the perfusion experimental chamber at room temperature and wash them by perfusion with standard solution for a few minutes before the beginning of the experiment.
3. Use the wavelengths described above (see *Protocol 2*) to image fura-2 responses in nucleoplasm. Use 488 nm argon laser line for excitation and emission $>$ 530 nm to measure responses of Calcium Green-1 dextran in nucleoplasm.
4. Use staining with ethidium bromide (excitation 363 nm, emission $>$ 520 nm) to confirm the nucleoplasmic localization of fluorescent dextrans (Plate 7d). We used 5 μg/ml of ethidium bromide in standard solution; the incubation takes approx. 5 min.

We have found that fluorescent dextrans accumulate in the internal area of the nuclei (Plate 7c). This allows us to monitor Ca^{2+} changes in the nucleoplasm. Ca^{2+} changes in the external solution resulted in rapid changes in free calcium in the nucleoplasm (*Figure 2A*). An external free calcium increase to 1 mM saturated the Ca^{2+} indicator in the nucleoplasm, whereas a subsequent application of 2 mM EGTA resulted in reversal of the signal. These data strongly suggest that the nuclear envelope is permeable to Ca^{2+}.

Stimulation of nuclei (pre-incubated with calcium in the presence of ATP) with IP_3 (*Figure 2B*) induced a rapid transient elevation of the nucleoplasmic calcium concentration. Stimulation of nuclei with cADP-ribose (*Figure 2C*) induced

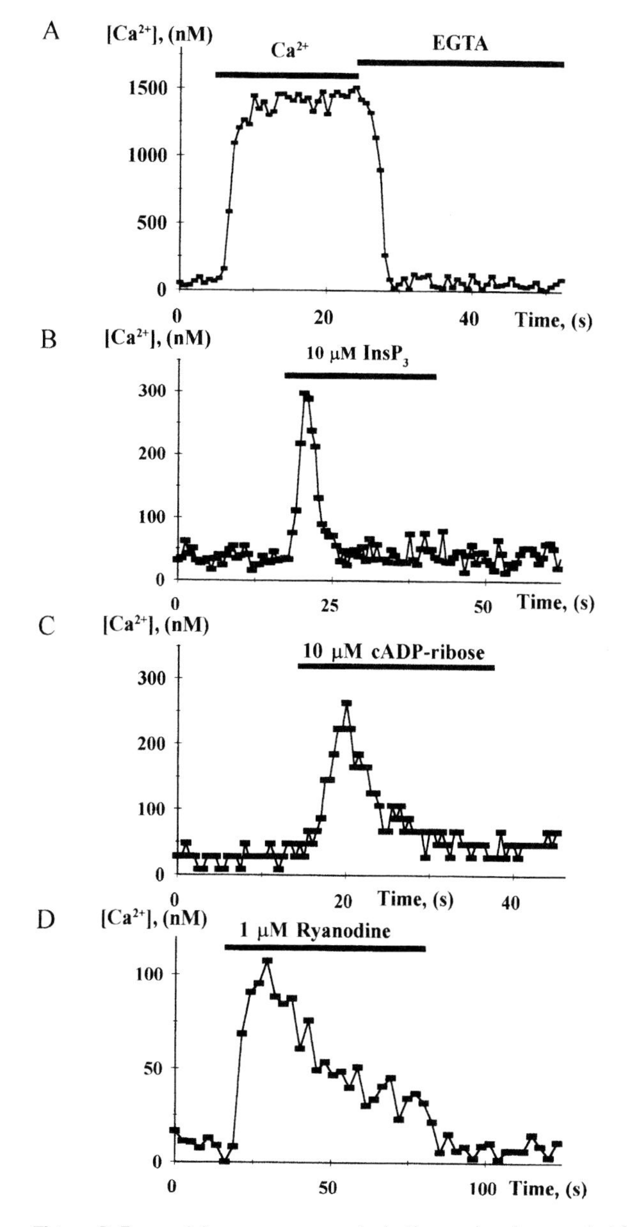

Figure 2 Free calcium measurements in the nucleoplasm using fura-2 dextran. The applications of 1 mM Ca^{2+} and 2 mM of EGTA to a single isolated nucleus demonstrate high permeability of the nuclear envelope. IP_3, cADP-ribose, and ryanodine induce Ca^{2+} releases apparently directed into the nucleoplasm. (A) Changes of external calcium concentration induce changes of nucleoplasmic calcium. (B) Application of IP_3 induced transient increase of free calcium in the nucleoplasm. (C) cADP-ribose induces nucleoplasmic calcium transient. (D) Ryanodine-induced calcium transient in the nucleoplasm. Reproduced with permission from Cell Press (1).

similar transient elevation of the free Ca^{2+} in the nucleoplasm measured with the dextran-bound Ca^{2+} indicators. The effect of cADP-ribose implies the presence of ryanodine receptors (25). Ryanodine itself evoked a rise in the nucleoplasmic calcium concentration (*Figure 2D*).

Ca^{2+} indicators attached to dextrans accumulated inside the nucleus and appear to record calcium release from the nuclear envelope, and remarkably this release is directed into the nucleoplasm. The transient nature of these responses can be explained by the leakage of calcium from the nucleoplasm through the nuclear pores.

3 Measurements of nuclear and cytosolic Ca^{2+} signals in intact isolated cells

This section describes the way we record, compare, and display calcium signals recorded in nucleoplasm and other regions of the cytosol of intact isolated cells.

The experiments described here show that it is possible to induce substantial Ca^{2+} signals in some regions of the cytosol without affecting the nucleus. It is also possible to elevate Ca^{2+} uniformly by sustained vigorous stimulations.

Plate 8 illustrates simultaneous confocal measurements of nuclear calcium and of calcium responses in two regions of cytosol (basal non-nuclear region and secretory region) of pancreatic acinar cells loaded with fura red. Cells were stained by incubation with 5 μM of fura red AM for 30 minutes at room temperature, and were then washed once by centrifugation. The fluorescence of fura red was excited by 488 nm argon laser line while the emission was recorded at above 515 nm. The positions of nuclei can be verified by staining with the specific nuclear dye Hoechst 33342 (Molecular Probes). We used the 363 nm argon laser line to excite Hoechst 33342. The emission was recorded with a barrier filter > 400 nm.

We used iontophoretic injection and pressure injection to stimulate the cells. Short iontophoretic application (about 3 sec) of ACh results in a Ca^{2+} increase in the secretory granule region, whereas no changes in cytoplasmic Ca^{2+} were detected in the basal area outside the nucleus or in the nucleus itself (please see Plates 8A, 8C). Plate 8Aa represents a transmitted non-confocal picture of a cluster of two pancreatic acinar cells. The dark areas in the cells correspond to secretory granule regions. Plate 8Ac–i shows the colour-coded changes of fluorescence intensities induced by a short application of ACh. Shade correction was applied to normalize a non-uniformity of indicator loading, by dividing all shown images by the image obtained a few seconds before the beginning of the transient. Plate 8Ac shows the distribution of the free cellular $[Ca^{2+}]$ before a short application of ACh which was started before Plate 8Ad was obtained.

During the short-lasting ACh application, the Ca^{2+} response develops in the secretory regions and disappears without spreading into the basal areas of the cells (please see Plate 8Ad–i). The $[Ca^{2+}]$ changes in the secretory granule regions, basal non-nuclear regions, and nuclear regions of the cell cluster were

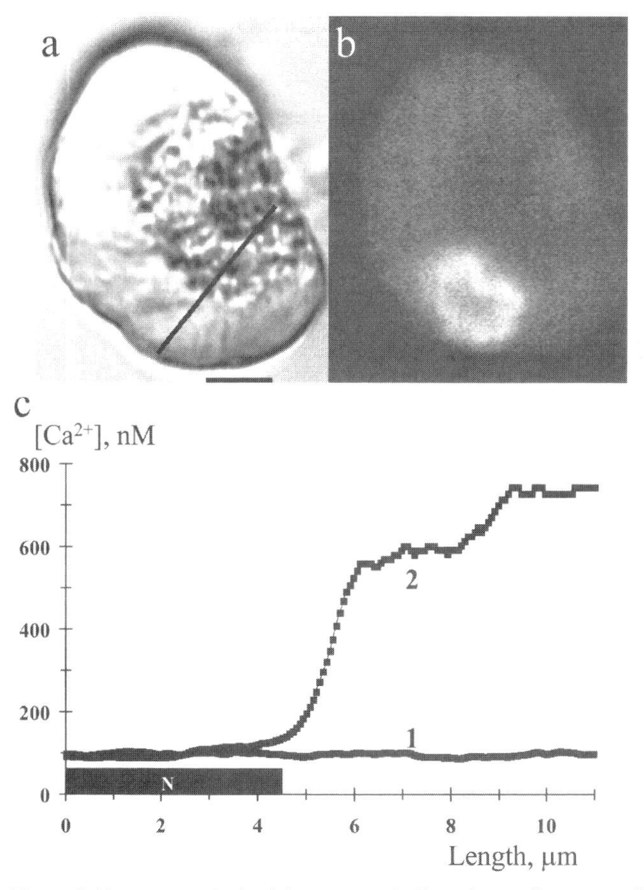

Figure 3 Measurement of calcium concentration along a line connecting the secretory granule region and the nucleus at rest and the peak of a localized Ca^{2+} response in the secretory region of the cell. The response was evoked by pressure application of ACh. (a) Transmitted light picture of pancreatic acinar cell. Scale bar corresponds to 5 μm. (b) The same pancreatic acinar cell after staining with Hoechst 33342 at the end of the experiment (to verify the position of the nucleus). (c) Ca^{2+} concentration along the line shown in (a), 1 at rest and 2 at the peak of the localized Ca^{2+} response. The bar labelled N represents the position of the nucleus. Reproduced with permission from Pflugers Archive—*Eur. J. Physiol.* (4).

measured (Plate 8A). The fluorescence of the nuclear region was higher than that of the cytoplasm. This non-uniformity of Ca^{2+} sensitive indicator distribution in the cells has been described before for other cell types (22). After each experiment cells were stained with Hoechst 33342 to verify the position of the nuclei (Plate 8Ab). It was found that at the peak of such localized Ca^{2+} responses, large Ca^{2+} gradients up to 400 nM/μm could be established along the connecting line between the secretory pole and nucleus (*Figure 3*). Longer applications of ACh (more than 10 sec) induced significant Ca^{2+} transients of similar amplitude in all three selected regions of the cell (please see Plates 8B, 8D).

Ionophoresis using double-barrelled microelectrode represents a convenient approach for application of ACh and controlling the duration of the stimulus. To generate a localized Ca^{2+} transient it was necessary to apply an ejecting current (5–20 nA) until the rise of intracellular Ca^{2+} started in the secretory granule region. Then the application of ACh was stopped by applying a retaining current (5–10 nA). This prevented the spreading of the Ca^{2+} signal into the basal part of the cell. The concentration of ACh in one barrel was 10 mM and another barrel was filled with 0.5 M of KCl. An ionophoresis injection system HVCS 02 (NPI, Germany) was used in our experiments.

Pressure application is another way of ensuring fast delivery of agonists. In our experiments short applications of ACh by pressure ejections induce localized Ca^{2+} transients in the secretory granule regions of the cells without spreading into the nuclei (*Figure 3*). Micropipettes for pressure ejection had approximately 1 μm tip diameter. ACh concentration in the micropipette solution was 10 μM (approximately ten times higher than supramaximal concentration of ACh for this cell type). For the application of ACh we used pressure pulses 0.03–0.1 bar delivered by Eppendorf microinjector 5242. The duration of pulses varied from 0.1–0.2 sec.

One of the advantages of pressure ejection is that the concentration of an agonist in cell vicinity can be easily estimated using a 'reporting' fluorescent indicator. Estimations of the ACh concentrations occurring in the vicinity of the cells as a result of short-lasting pressure ejections were made using pressure applications of a fluorescent indicator (rhodamine). The changes of rhodamine fluorescence were then assessed using the confocal microscope in high confocality mode. The resulting changes of fluorescence (*Figure 4*) were compared with the fluorescence recorded from small droplets of solution with 0.1, 1, and 10 μM of rhodamine. The concentration of rhodamine in the pipette was 10 μM. After approximately 0.5 sec from the moment of ejection the concentration declines to half the original value and becomes indistinguishable from the level

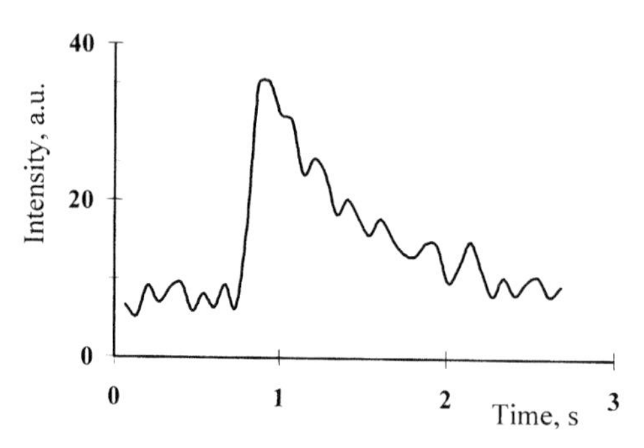

Figure 4 Changes of rhodamine B fluorescence after pressure application. Rhodamine was injected by pressure and the changes of fluorescence were recorded at a distance from the pipette that corresponds to the distance between the pipette and the centre of the cell in experiments with pressure application of ACh (approximately 10 μM). Pressure applied in this experiment was 0.04 bar. The peak fluorescence of rhodamine was approximately nine times smaller than the fluorescence measured from a droplet containing 10 μM rhodamine (concentration of rhodamine in the pipette). Reproduced with permission from Pflugers Archive – Eur. J. Physiol. (4).

before ejection after 1.5 sec (*Figure 4*). The results of these measurements were used to estimate the ACh concentration (the estimated peak ACh concentration in the vicinity of the cell was in the region of 0.6–1.7 μM). These results indicate that a short (~ 1 sec) relatively high concentration of ACh application results in localized cytosolic Ca^{2+} spikes in the secretory pole.

References

1. Gerasimenko, O. V., Gerasimenko, J. V., Tepikin, A. V., and Petersen, O. H. (1995). *Cell*, **80**, 439.
2. Gerasimenko, O. V., Gerasimenko, J. V., Tepikin, A. V., and Petersen, O. H. (1996). *Pflugers Arch.*, **432**, 1.
3. Petresen, O. H., Gerasimenko, O. V., Gerasimenko, J. V., Mogami, H., and Tepikin, A. V. (1996). *Cell Calcium*, **23**, 87.
4. Gerasimenko, O. V., Gerasimenko, J. V., Tepikin, A. V., and Petersen, O. H. (1996). *Pflugers Arch.*, **432**, 1055.
5. Berridge, M. J. (1993). *Nature*, **361**, 315.
6. Petersen, O. H., Petersen, C. C. H., and Kasai, H. (1994). *Annu. Rev. Physiol.*, **56**, 297.
7. Bootman, M. D. and Berridge, M. J. (1995). *Cell*, **83**, 675.
8. Newport, J. and Forbes, D. J. (1987). *Annu. Rev. Biochem.*, **56**, 535.
9. Pante, N. and Aebi, U. (1994). *Curr. Opin. Struct. Biol.*, **4**, 187.
10. Sweet, D. J. and Gerace, L. (1995). *Trends Cell Biol.*, **5**, 444.
11. Lang, I., Schulz, M., and Peters, R. (1986). *J. Cell Biol.*, **102**, 1183.
12. Tombes, R. M., Simerly, C., Borisy, G. G., and Schatten, G. (1992). *J. Cell Biol.*, **117**, 799.
13. Lanini, L., Bachs, O., and Carafoli, E. (1992). *J. Biol. Chem.*, **267**, 11548.
14. Serratosa, J., Pujol, M. J., Bachs, O., and Carafoli, E. (1988). *Biochem. Biophys. Res. Commun.*, **142**, 1162.
15. Burns, K., Duggan, B., Atkinson, E. A., Famulski, K. S., Nemer, M., Bleackley, R. C., *et al.* (1994). *Nature*, **367**, 476.
16. Nicotera, P., Orrenius, S., Nilsson, T., and Berggren, P.-O. (1990). *Proc. Natl. Acad. Sci. USA*, **87**, 6858.
17. Malviya, A. N., Rogue, P., and Vincendon, G. (1990). *Proc. Natl. Acad. Sci. USA*, **87**, 9270.
18. Kume, S., Muto, A., Aruya, J., Nakagawa, T., Michikawa, T., Furuiche, T., *et al.* (1993). *Cell*, **73**, 555.
19. Divecha, N., Banfic, H., and Irvine, R. F. (1993). *Cell*, **74**, 405.
20. Gillot, I. and Whitaker, M. (1993). *J. Exp. Biol.*, **184**, 213.
21. Brini, M., Murgia, M., Pasti, L., Picard, D., Pozzan, T., and Rizzuto, R. (1993). *EMBO J.*, **12**, 4813.
22. Al-Mohanna, F. A., Caddy, K. W. T., and Bolsover, S. R. (1994). *Nature*, **367**, 745.
23. Kasai, H., Li, Y. X., and Miyashita, Y. (1993). *Cell*, **74**, 669.
24. Maruyama, Y. and Petersen, O. H. (1994). *Cell Calcium*, **16**, 419.
25. Nicotera, P., McConkey, D. J., Jones, D. P., and Orrenius, S. (1989). *Proc. Natl. Acad. Sci. USA*, **86**, 453.
26. Hechtenberg, S. and Beyersmann, D. (1993). *Biochem. J.*, **289**, 757.
27. Thorn, P., Lawrie, A. M., Smith, P. M., Gallacher, D. V., and Petersen, O. H. (1993). *Cell*, **74**, 661.
28. Lee, H. C., Walseth, T. F., Bratt, G. T., Hayes, R. N., and Clapper, D. L. (1989). *J. Biol. Chem.*, **264**, 1608.
29. Hua, S.-Y., Tokimasa, T., Takasawa, S., Furuya, Y., Nohmi, M., Okamoto, H., *et al.* (1994). *Neuron*, **12**, 1073.
30. Thorn, P., Gerasimenko, O., and Petersen, O. H. (1994). *EMBO J.*, **13**, 2038.

Part Three

Monitoring specific calcium reactions

Chapter 8

Controlling cytoplasmic calcium and measuring calcium-dependent gene expression in intact cells

Ricardo E. Dolmetsch
Division of Neuroscience Children's Hospital, Harvard Medical School, 300 Longwood Avenue, Boston, MA 02115, USA.

Paul A. Negulescu
Aurora Biosciences, 11010 Torreyana Road, San Diego, CA 92121, USA.

1 Introduction

Ca^{2+}-regulated gene expression plays an important role in many cellular events including proliferation, survival, and maturation. Ca^{2+} controls gene expression both by binding directly to transcription factors and by influencing signal transduction pathways that regulate transcription factors (1). Despite the long and growing list of Ca^{2+}-regulated genes, relatively little is known about how these genes are affected by the temporal and spatial patterns of cytoplasmic calcium signals (2). The aim of this chapter is to provide a practical guide to some of the methods that have been recently developed to study Ca^{2+}-dependent gene expression in response to cytoplasmic Ca^{2+} ($[Ca^{2+}]_i$) signals. It is divided into two sections; the first deals with the use of reporter genes for measuring gene expression in single cells and the second with methods for generating defined $[Ca^{2+}]_i$ signals in either single cells or in populations of cells. These two techniques have proved to be valuable in analysing the effects of $[Ca^{2+}]_i$ patterns on the activation of transcription factors and on the induction of genes.

Ca^{2+} can be measured in single cells with high spatial and temporal resolution by using indicator dyes and protein sensors. Experiments using these reagents in conjunction with video microscopy have revealed a large diversity of $[Ca^{2+}]_i$ responses in single cells (3). $[Ca^{2+}]_i$ oscillations and waves as well as plateaus and isolated spikes have been observed in nearly every cell type which has been studied. Three general features of these responses make studying the downstream effects of $[Ca^{2+}]_i$ signals particularly challenging. First, $[Ca^{2+}]_i$ responses are heterogeneous among cells in a population so that only a fraction display a particular signalling pattern (4). Secondly, the signals in single cells are variable as a function of time so that a cell may display a whole spectrum of signalling patterns if observed over a period of hours. Finally, calcium signals are often triggered by stimulation of cell surface receptors which are linked to signalling pathways other than calcium. The temporal and spatial variability of $[Ca^{2+}]_i$ responses makes it difficult to correlate calcium patterns with the expression of specific genes in cell populations. While the pleiotropy of receptors complicates the isolation of the gene regulatory effects of Ca^{2+} from the effects of other signalling pathways.

Two general approaches can be used to circumvent these problems. First, gene expression can be measured at the single cell level in intact cells. Single cell measurement of genes allows the experimenter to correlate $[Ca^{2+}]_i$ patterns measured in an individual cell with the induction of a gene in that same cell. This eliminates the uncertainty associated with measuring gene expression in a population of cells with diverse $[Ca^{2+}]_i$ signalling patterns. A second approach is to synchronize the $[Ca^{2+}]_i$ signals in a large population of cells. This allows an investigator to study biochemical events in a population of cells with a uniform calcium history. This second approach has the additional advantage that it allows $[Ca^{2+}]_i$ to be studied independently of other signalling pathways activated by cell surface receptors. These techniques taken together comprise a powerful approach to investigating Ca^{2+}-dependent gene expression in cells.

2 Measuring gene expression in single cells

Several methods exist for measuring gene expression at the single cell level, each with its own set of advantages and disadvantages. For example, protein measurement techniques such as immunocytochemistry (e.g. fluorescently tagged antibodies) are useful because they measure the expression of an endogenous protein and reflect the activities of endogenous promoters and enhancers in the appropriate chromosomal context. However, specific antibodies to various proteins are not always available. Furthermore, these techniques are relatively insensitive and are cumbersome to routinely perform on a microscope stage. Single cell PCR can be highly sensitive, but is difficult to perform on more than a few cells and may be subject to artefacts resulting from the high number of PCR cycles needed to obtain a signal.

2.1 Reporter genes

In the past several years there has been a considerable increase in the use of reporter genes as markers for gene regulation (5). These reporters are usually enzymes, that can be engineered downstream of a promoter of interest, expressed in cells, and monitored using appropriate substrates. This approach has the advantage of providing large amplification of small signals as well as a comparatively simple and quantitative way of assessing the expression of a specific

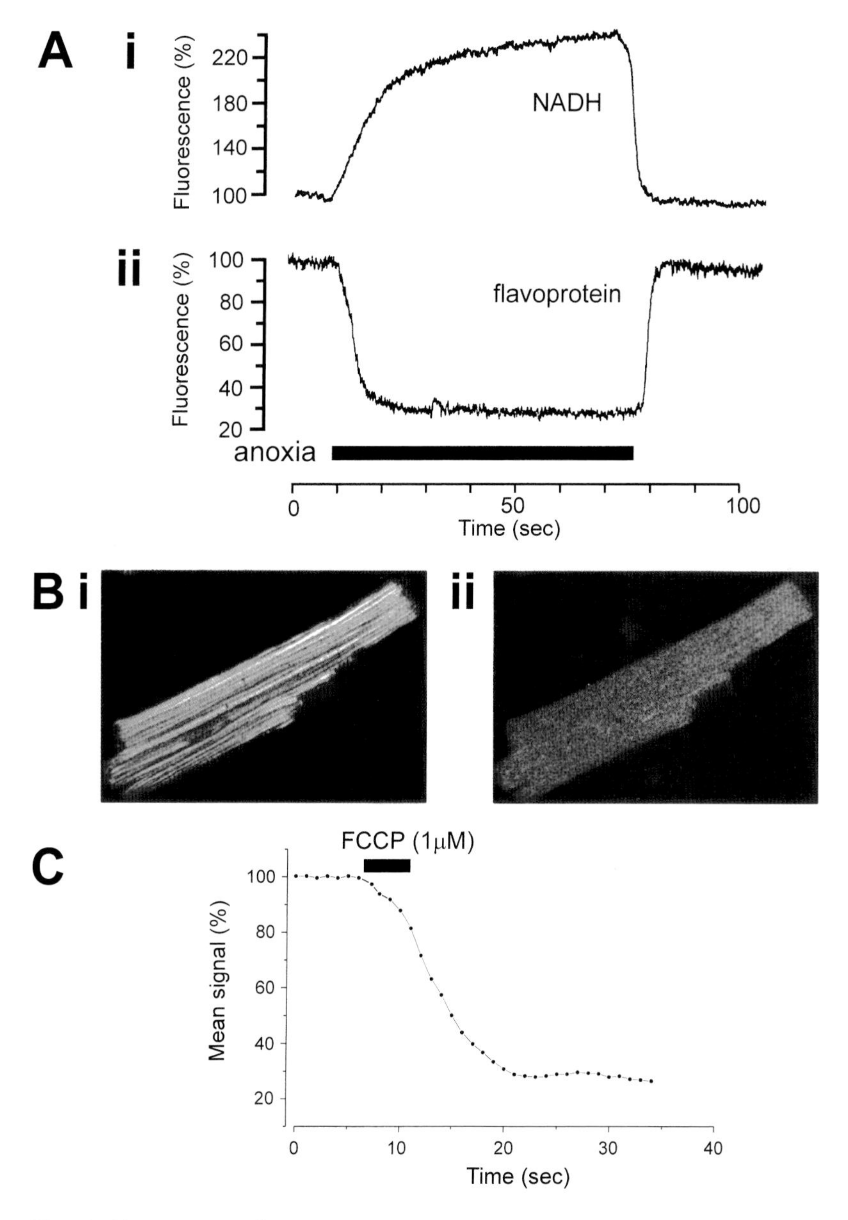

Plate 1 Changes in autofluorescence signals. (A) Changes in (i) NADH and (ii) flavoprotein autofluorescence in response to anoxia. These records come from a rat sensory neuron, and show the changes in signal recorded during superfusion of the cell with saline equilibrated with nitrogen with the addition of 500 μM dithionite. NADH autofluorescence was excited at 350 nm and measured at 450 nm $\pm$ 40 nm, and flavoprotein autofluorescence was excited at 450 nm and measured at 550 nm $\pm$ 40 nm. (B) Confocal images of NADH autofluorescence in freshly dissociated rat cardiomyocytes. The cells were illuminated with the 351 nm laser line (at just 2% of full power) of a Zeiss LSM510, and light was collected between 380 and 470 nm—not perhaps the most efficient configuration, but a default filter set available on the microscope. In (i) the appearance of mitochondria, characteristically arranged in rows along the axis of the cell, is evident. FCCP (1 μM) was applied briefly from a nearby puffer pipette, and the signal largely disappeared, leaving only the diffuse blue fluorescence shown in (ii). The plot in (C) shows the time course of the response.

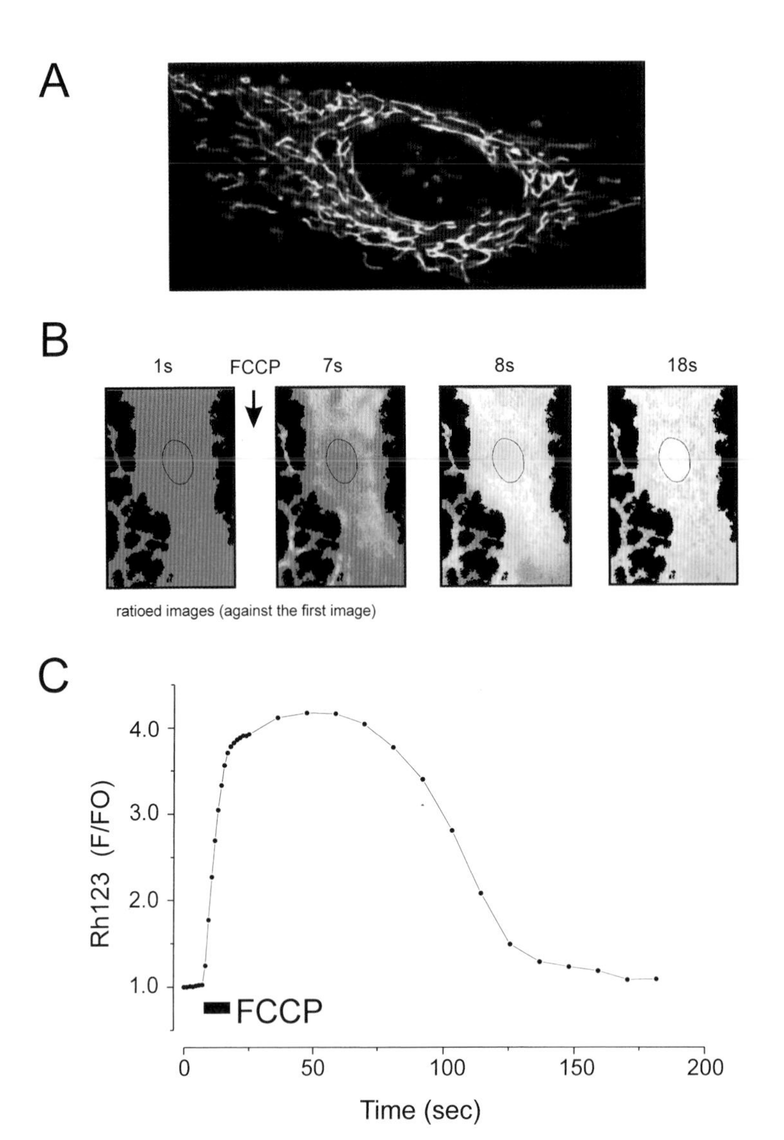

Plate 2 Imaging mitochondrial membrane potential ($\Delta\psi_m$) in the 'dequench' mode. (A) An image of a rat cortical astrocyte stained with tetramethyl-rhodamine ethyl ester (TMRE) at 3 μM for 15 minutes followed by washing. The image was obtained on a confocal microscope (Zeiss LSM510) illuminating at 488 nm and collecting the signal with a long pass filter $>$ 560 nm. The mitochondrial distribution—and the configuration of mitochondria in these cells—is evident. (B) A series of images, also of an astrocyte loaded similarly with TMRE, taken from a time sequence to illustrate the changes in signal following the collapse of $\Delta\psi_m$ after brief application of FCCP (1 μM) from a puffer pipette. In this instance, the images have been processed by constructing a ratio with respect to the first image of the series, so that each pixel shows the proportional change in signal. Note that immediately after FCCP application, an increase in fluorescence can be seen close to the mitochondria around the nucleus, gradually diffusing through the cell, and finally accumulating throughout the cytosol and in the nucleus. The mitochondria can now be seen as dark shadows against the brighter background, as here the signal has changed little, while it has increased throughout the cell. The images were obtained using a cooled CCD camera (Hamamatsu 4880). (C) A plot of intensity with time from a very similar experiment but using the dye Rh123, loaded at 10 μg/ml for 10 minutes followed by washing.

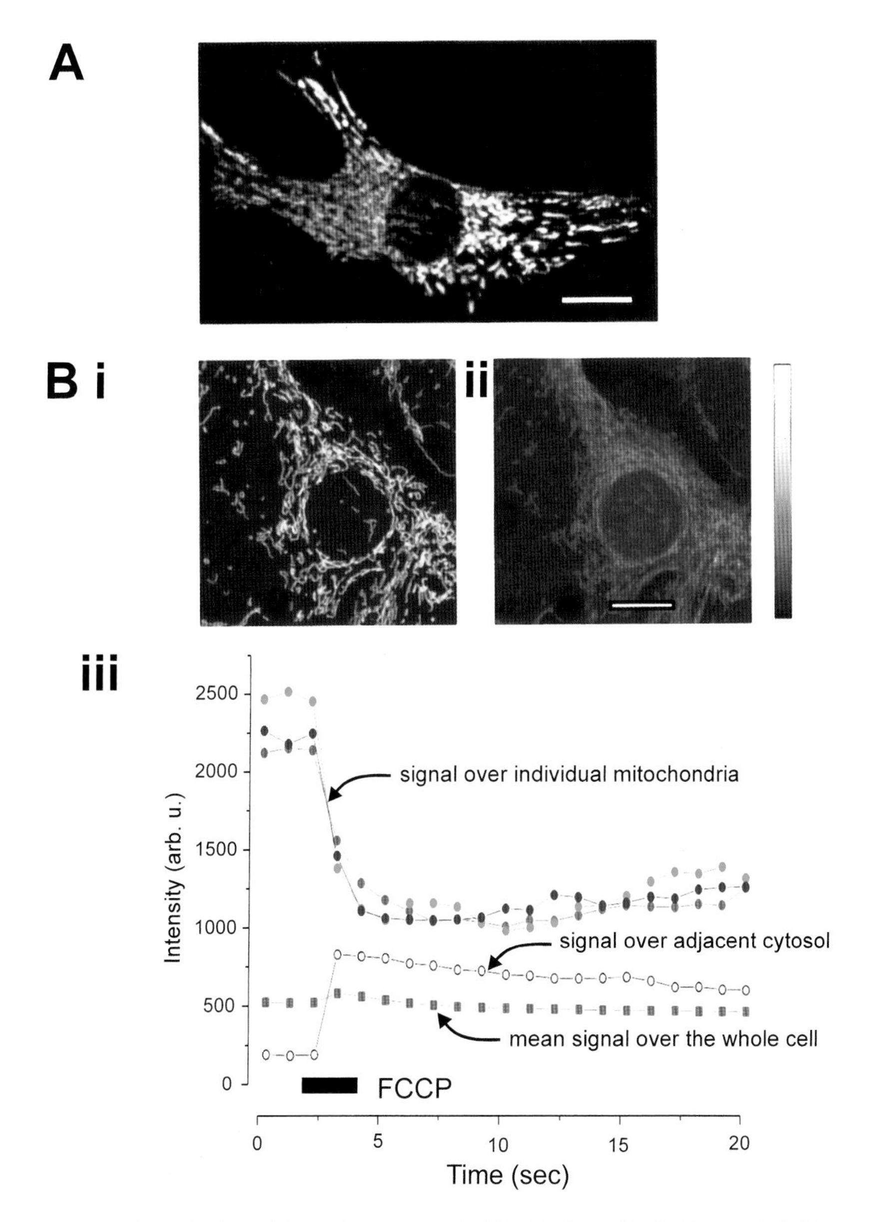

Plate 3 Imaging mitochondrial membrane potential ($\Delta\psi_m$) in the redistribution mode. (A) An image of a neonatal rat cardiomyocyte, and (Bi) a rat cortical astrocyte in culture in which TMRE was added to the bathing saline at a concentration of 15 nM and allowed to equilibrate for ~ 20 minutes. (Bii) After application of 1 μM FCCP to collapse $\Delta\psi_m$ the dye has shown a redistribution throughout the cytosol. The display is 'mapped' non-linearly (see scale bar) in order to display both signals, so great is the apparent loss of signal from the mitochondria. However, measurement of the average signal from the whole cell barely changed at all. This is shown in (C) in which the intensity is plotted over three mitochondria (closed circles), from an area of cytosol adjacent to some mitochondria (open circles), over which the signal increased, and averaged over the whole cell (filled squares), over which the signal did not change significantly. Clearly, the signal measured from the whole cell would be uninterpretable. In this instance, images were obtained on a confocal system as described in Plate 2A.

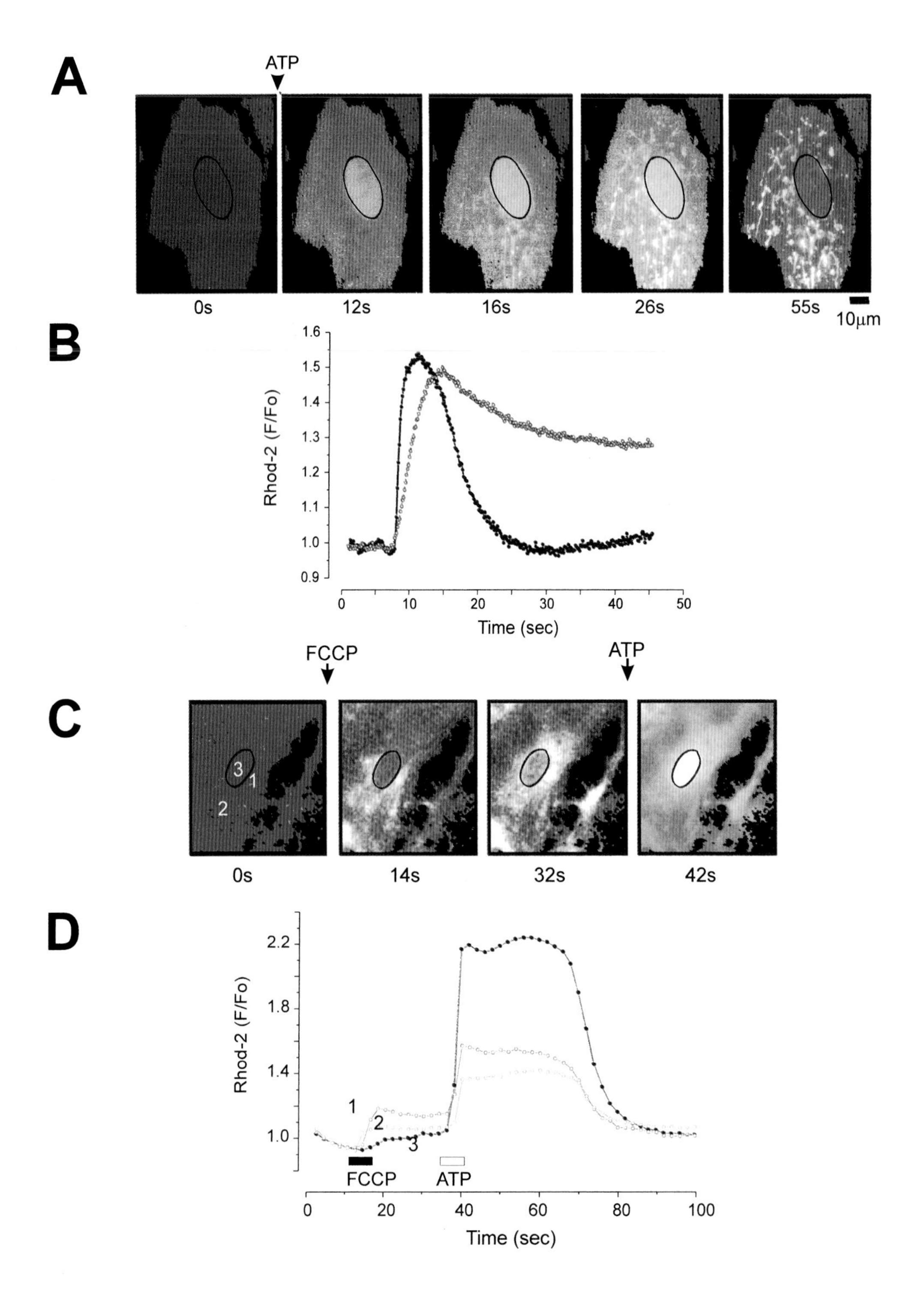
A
ATP
0s
12s
16s
26s
55s
10μm
B
Rhod-2 (F/Fo)
Time (sec)
C
FCCP
ATP
3
1
2
0s
14s
32s
42s
D
Rhod-2 (F/Fo)
1
2
3
FCCP
ATP
Time (sec)

Plate 4 Measurements of mitochondrial Ca^{2+} uptake using rhod-2. (A) A series of images of a rat cortical astrocyte loaded with rhod-2 (5 μM for 30 minutes) selected from a time series to show the response to 100 μM ATP. The images were acquired on a cooled CCD camera (Digital Pixel frame transfer) and have been ratioed with respect to the first image to enhance the changes in fluorescence. ATP application mobilizes $[Ca^{2+}]_i$ from ER stores and initially caused a diffuse increase in cytosolic signal, best seen over the nucleus (indicated by circle). Gradually, the relative signal over the mitochondria increased, starting at the lower pole of the cell, and then spreading throughout the cell. The mitochondrial signal stayed bright long after the cytosolic signal had recovered. (B) A plot, obtained using the faster read-out Hamamatsu 4880 camera, to show the change in signal over a mitochondrion (open circles) and over the nucleus (filled circles, representing the cytosolic, non-mitochondrial signal). In these very flat cells, mitochondrial structure can be imaged without needing confocal microscopy, but equivalent experiments using neurons absolutely require confocal imaging as the mitochondrial signal tends to be swamped by the (low concentration, but greater volume of) residual dye in the cytosol. (C and D) illustrate the dependence of mitochondrial Ca^{2+} uptake on $\Delta\psi_m$ and also help to validate the interpretation of (A and B). Again, the images have been ratioed with respect to the first of the series. By this processing, mitochondrial rhod-2 fluorescence in the first frame is removed to enhance the changes in the cytosolic fluorescence. Application of FCCP (1 μM) in the presence of oligomycin (2.5 μg/ml) induced small 'puffs' of fluorescence (marked 1 and 2) around the mitochondria, suggesting some small release of Ca^{2+} in response to the mitochondrial depolarization. Application of ATP then produced a large global increase in signal showing the same time course over cytosol and over the nucleus (encircled and marked 3), without the appearance of punctate fluorescence as in (A), in the absence of a sustained mitochondrial response. (D) The signal intensities over the nucleus and the cytosol near mitochondria, are plotted as a function of time.

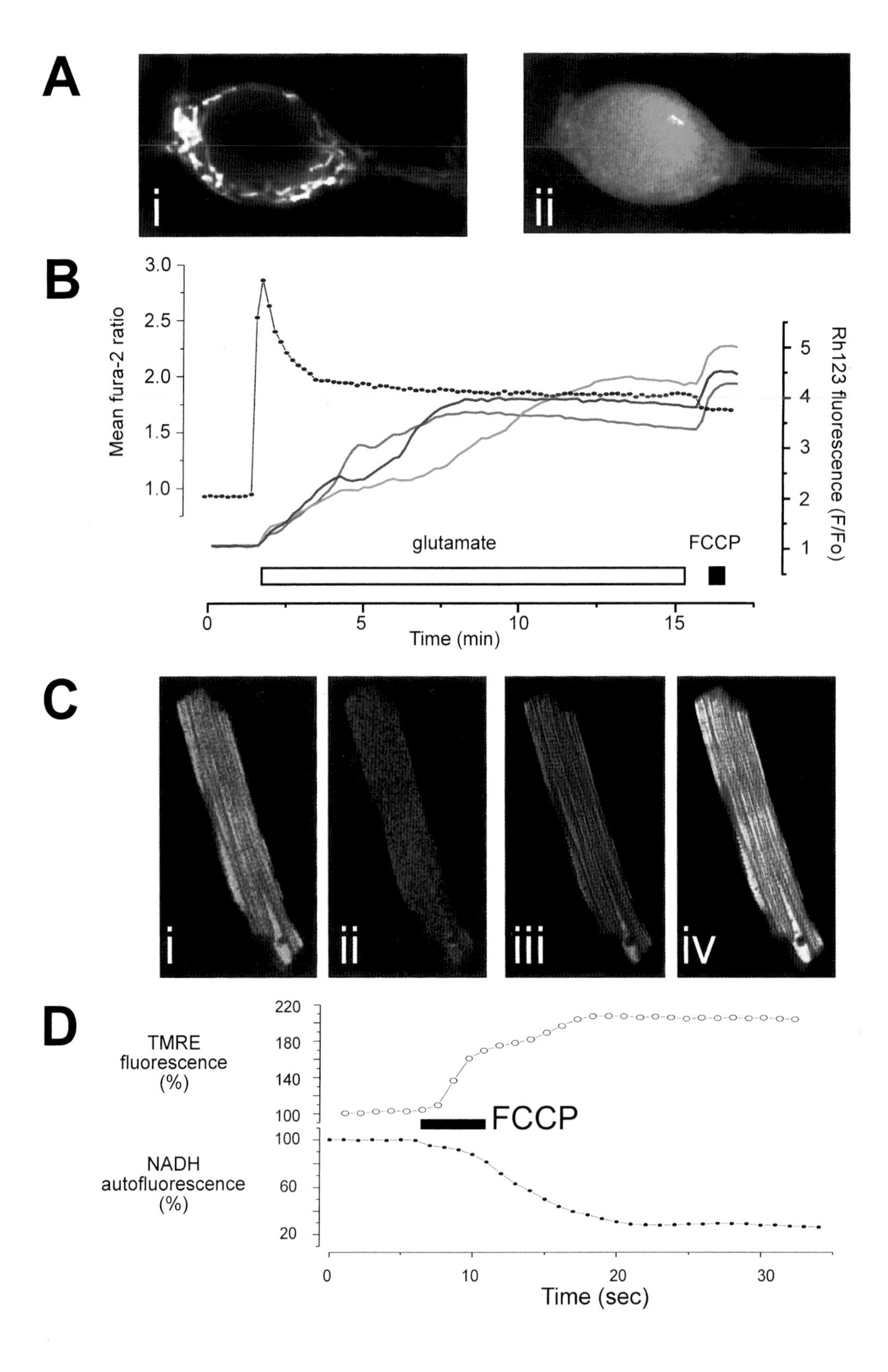
A
i
ii
B
Mean fura-2 ratio
3.0
2.5
2.0
1.5
1.0
Rh123 fluorescence (F/Fo)
5
4
3
2
1
glutamate
FCCP
0
5
10
15
Time (min)
C
i
ii
iii
iv
D
TMRE fluorescence (%)
220
180
140
100
FCCP
NADH autofluorescence (%)
100
60
20
0
10
20
30
Time (sec)

Plate 5 Simultaneous measurements of mitochondrial and other variables. (Ai and ii) Images of a hippocampal neuron loaded with both fura-2 and Rh123, and imaged on a confocal microscope (Zeiss LSM 510). The images were obtained after filtering at 505–550 nm, and the excitation was switched from 488 nm (i, Rh123) to 351 nm (ii, fura-2), illustrating the selective excitation of each indicator at each wavelength. (B) Simultaneous measurements of $[Ca^{2+}]_i$ (filled circles, using fura-2) and $\Delta\psi_m$ using Rh123 (solid traces) in rat hippocampal neurons in culture viewed at low power (x 20 objective). Application of glutamate at toxic levels (100 μM with glycine 10 μM) raised $[Ca^{2+}]_{cyt}$ in all cells, and was followed by a gradual loss of $\Delta\psi_m$ that varied in time and amplitude between neurons in the field. The images were acquired on a cooled CCD using a stepping filter wheel to step sequentially between 340, 380, and 490 nm. Light was collected using a long pass filter $>$ 510 nm. (C) NADH autofluorescence and $\Delta\psi_m$ were measured simultaneously from a rat cardiomyocyte loaded with TMRE (3 μM for 15 minutes followed by washing). The images were obtained using a confocal system, with simultaneous excitation at both 351 nm and 543 nm using a multiple band dichroic mirror, and measuring emitted signals on two channels, with one filter set at 380–470 nm and the other at $>$ 560. Brief application of 1 μM FCCP caused a loss of the autofluorescence (i and ii) and a dequench (iii and iv, an increase) in the TMRE signal, plotted as a function of time (D).

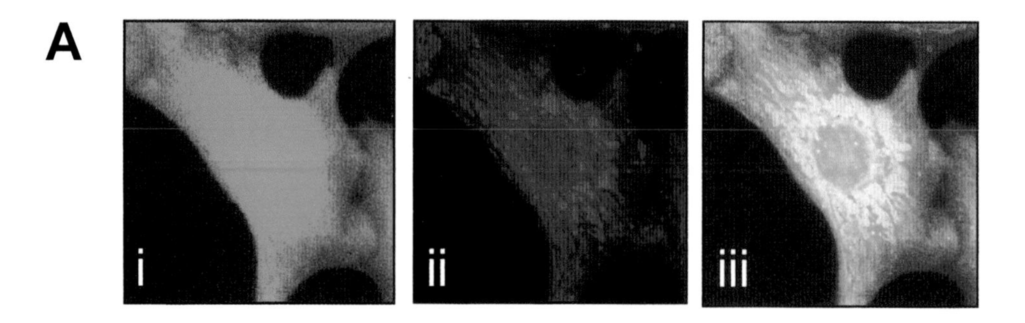

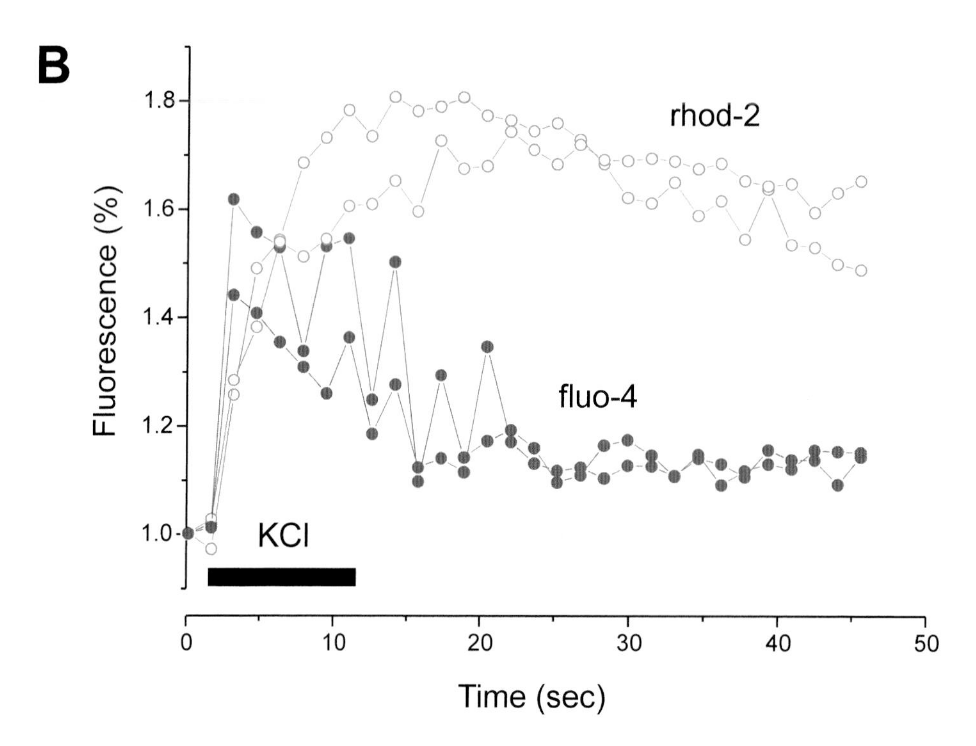

Plate 6 (A) Images of cytosolic [Ca^{2+}] and mitochondrial [Ca^{2+}] using (i) fluo-4 and (ii) rhod-2 in a neonatal cardiomyocyte in culture. Image (iii) shows the overlay of the two channels. (B) Membrane depolarization with 50 mM KCl raised $[Ca^{2+}]_{cyt}$ (filled circles) and mitochondrial [Ca^{2+}] (open circles) and the time course and relative amplitudes of the signals originating from each dye plotted from an area immediately over several mitochondria were clearly quite different. These images were acquired on a confocal system, excited at 488 nm, and measured on two channels with one filter set at 505–550 nm (fluo-4) and the other to $>$ 560 nm (rhod-2).

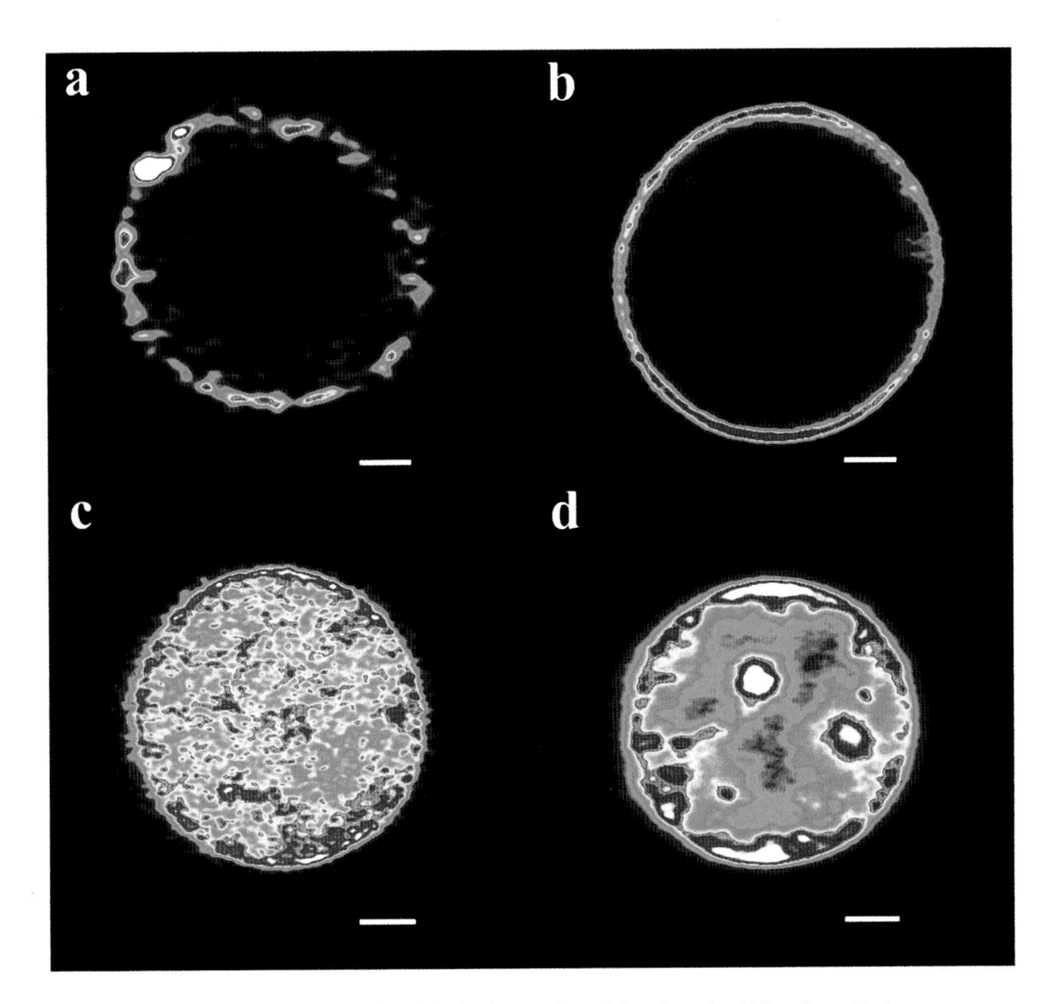

Plate 7 The distribution of fura-2, $DiOC_6(3)$, dextran-bound fura-2, and ethidium bromide in isolated nuclei. (a) Fluorescence of fura-2 AM loaded nucleus (excitation at 363 nm—close to the isobestic point, emission at 505 nm). (b) Fluorescence of $DiOC_6(3)$ loaded nucleus (excitation at 488 nm, emission at 530 nm). (c) Fluorescence of fura-2 dextran incubated nucleus (excitation at 363 nm—close to the isobestic point, emission at 505 nm). (d) Fluorescence of ethidium bromide stained nucleus (excitation at 363 nm, emission at 530 nm). Reproduced with permission from Cell Press (1).

Plate 8 Ca^{2+} signals in different regions of pancreatic acinar cells stimulated by acetylcholine (ACh). Confocal images of the cytosolic calcium signal after short application of ACh. (Aa) Transmitted light picture of cluster of two pancreatic acinar cells. The regions of interest are shown on this picture: yellow boxes, secretory granule regions; blue boxes, the nuclei; red, the non-nuclear, basal pole of the cells. Scale bar corresponds to 5 μm. (Ab) The same cell cluster incubated with Hoechst 33342 at the end of experiment to verify the position of the nuclei. (Ac) Distribution of fura red fluorescent intensity before stimulation. ACh was applied before (d). (Ad–i) Changes in fluorescence induced by short (3 sec) application of ACh (shade corrected images). Intervals between images were 0.53 sec. An inverted linear colour scale was used: red coloration corresponds to low fluorescence (high [Ca^{2+}]), blue coloration to high fluorescence (low [Ca^{2+}]). (B) Confocal images of cytosolic calcium signals after long (10 sec) application of ACh (shade corrected images, time interval between images was 1.06 sec). Scale bar corresponds to 5 μm. The images correspond to experiments shown in (D) and from the same cell cluster illustrated in (A). Images (Ba) and (Bb) were obtained before ACh addition whereas images (Bc–i) were obtained during ACh exposure. (C) Changes of calcium concentration in selected regions of interest (shown in A) after short ACh stimulation. Bars represent time of ACh application. (D) Changes of calcium concentration in selected regions of interest (shown in A) after long-lasting ACh application. Bars represent time of ACh stimulation. Reproduced with permission from Pflugers Archive – Eur. J. Physiol. (4).

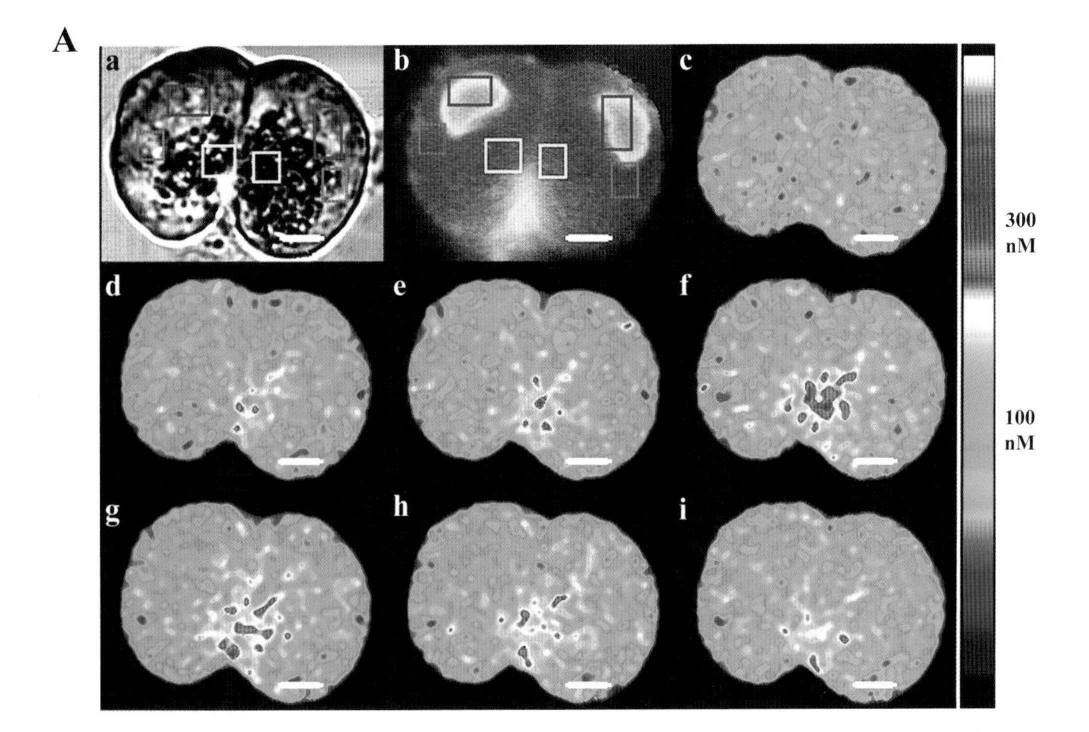
A
a
b
c
d
e
f
g
h
i
300
nM
100
nM

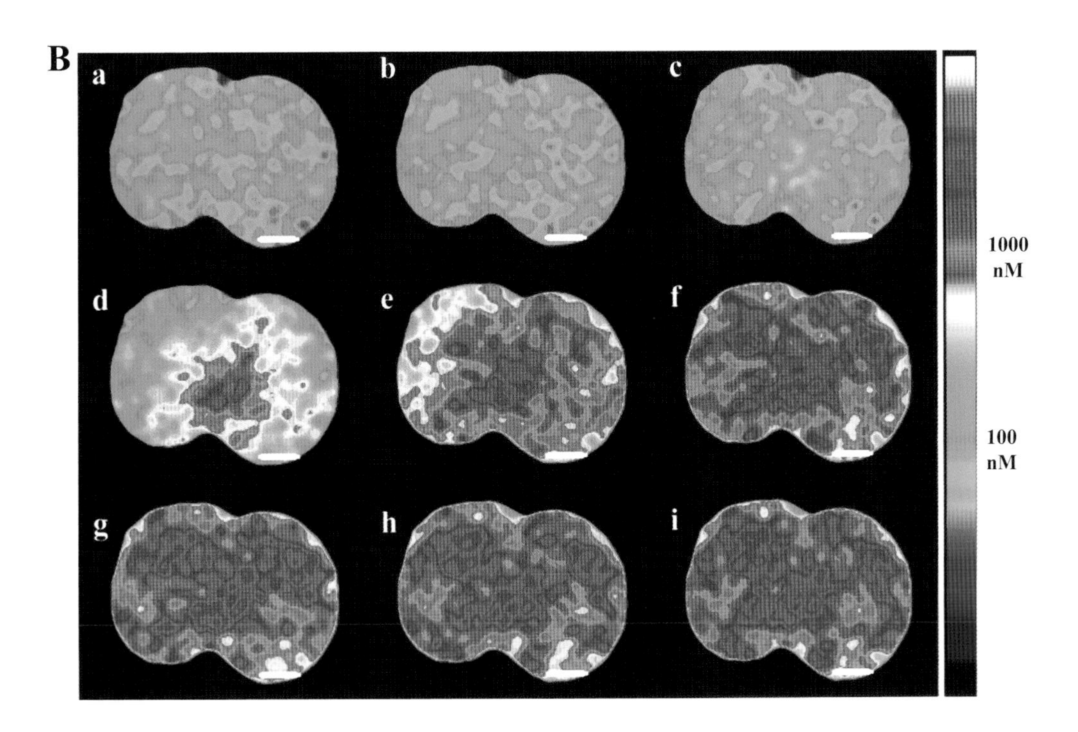
B
a
b
c
d
e
f
g
h
i
1000
nM
100
nM

Left cell

Right cell

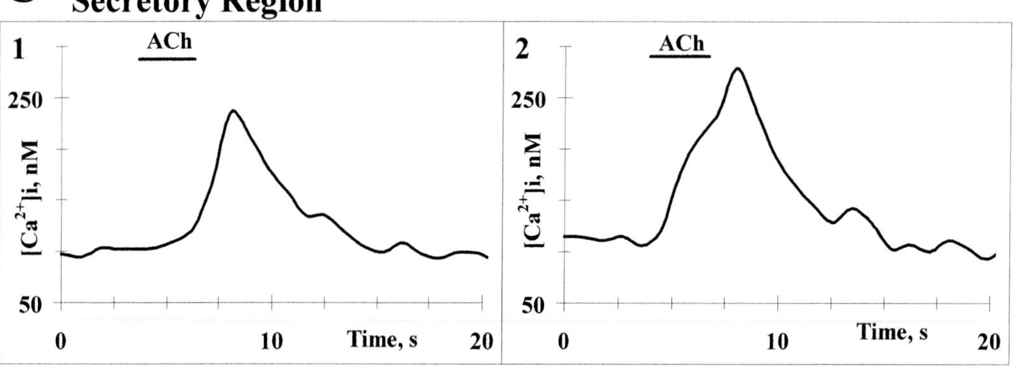

Basal Region

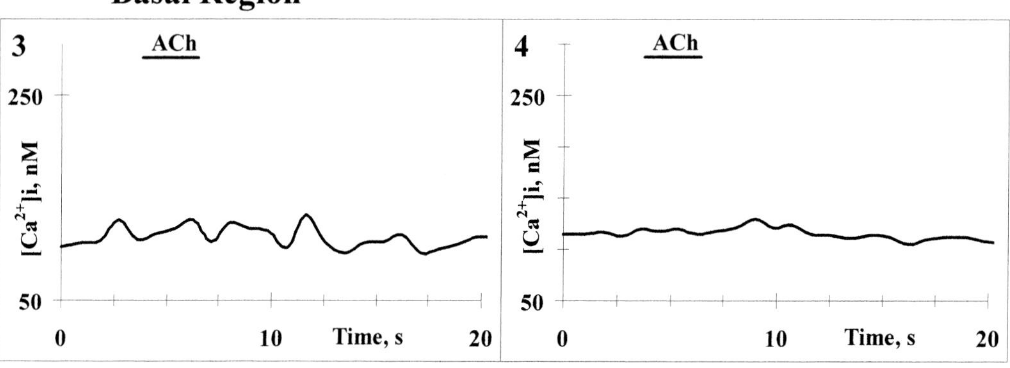

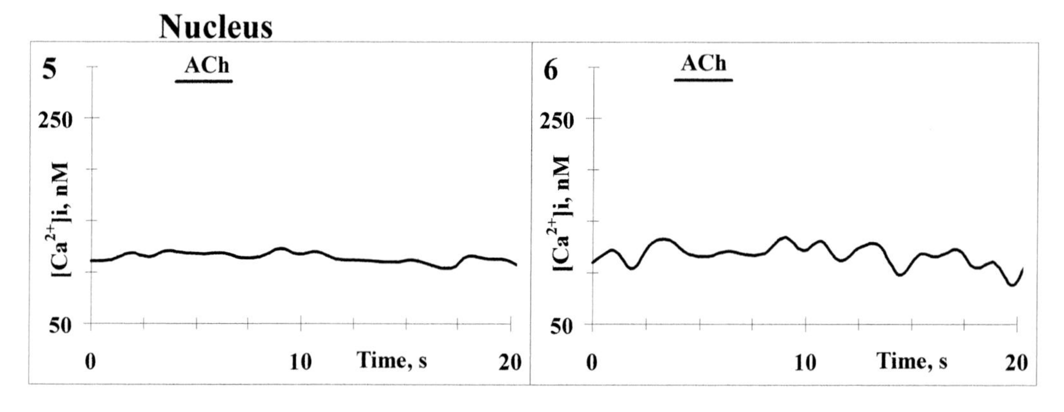

Plate 8 *(continued)*

D **Left cell** **Right cell**

Secretory Region

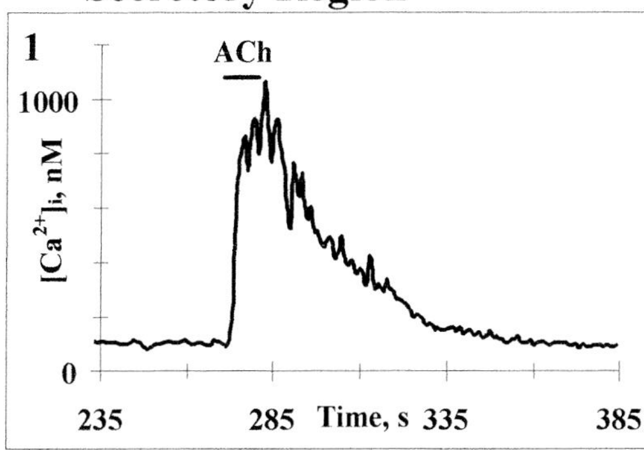

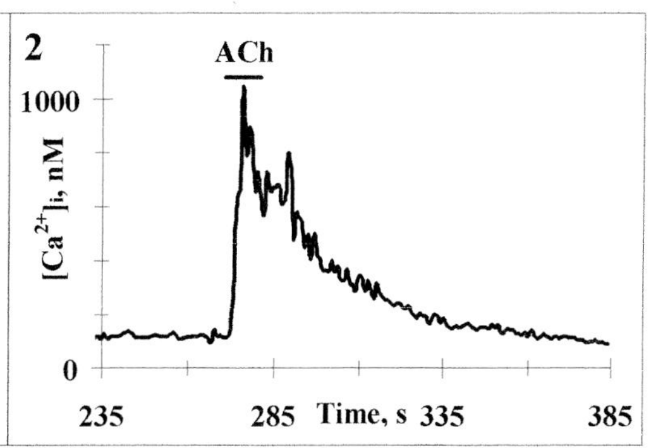

Basal Region

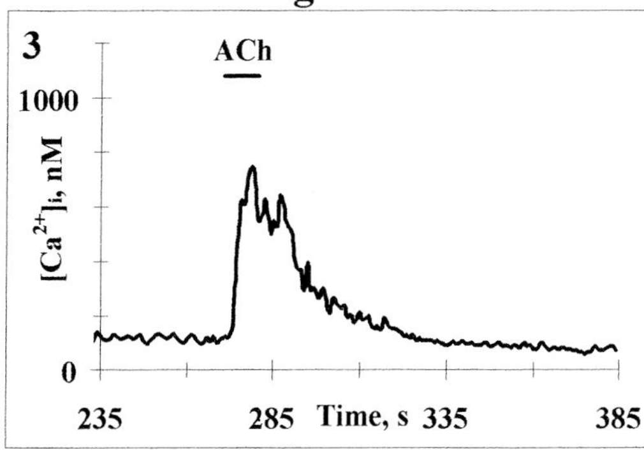

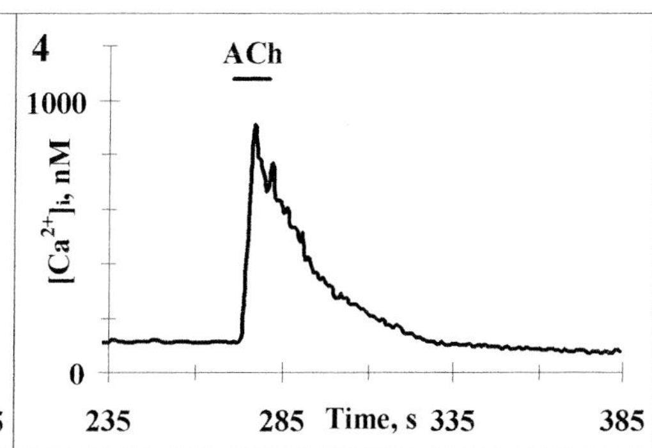

Nucleus

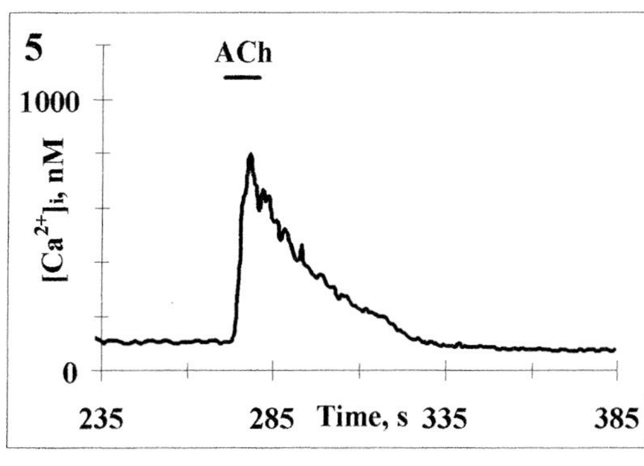

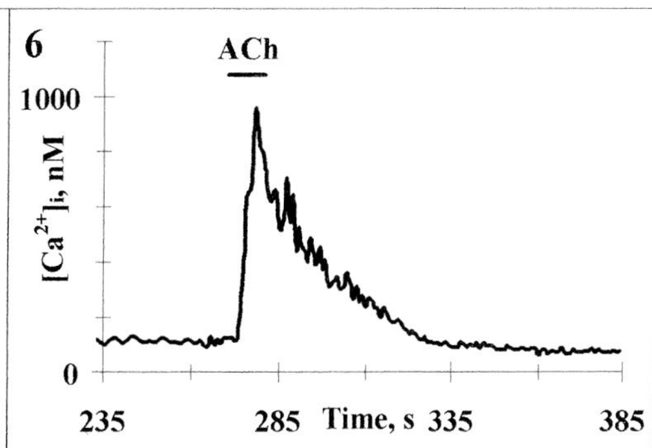

Plate 8 *(continued)*

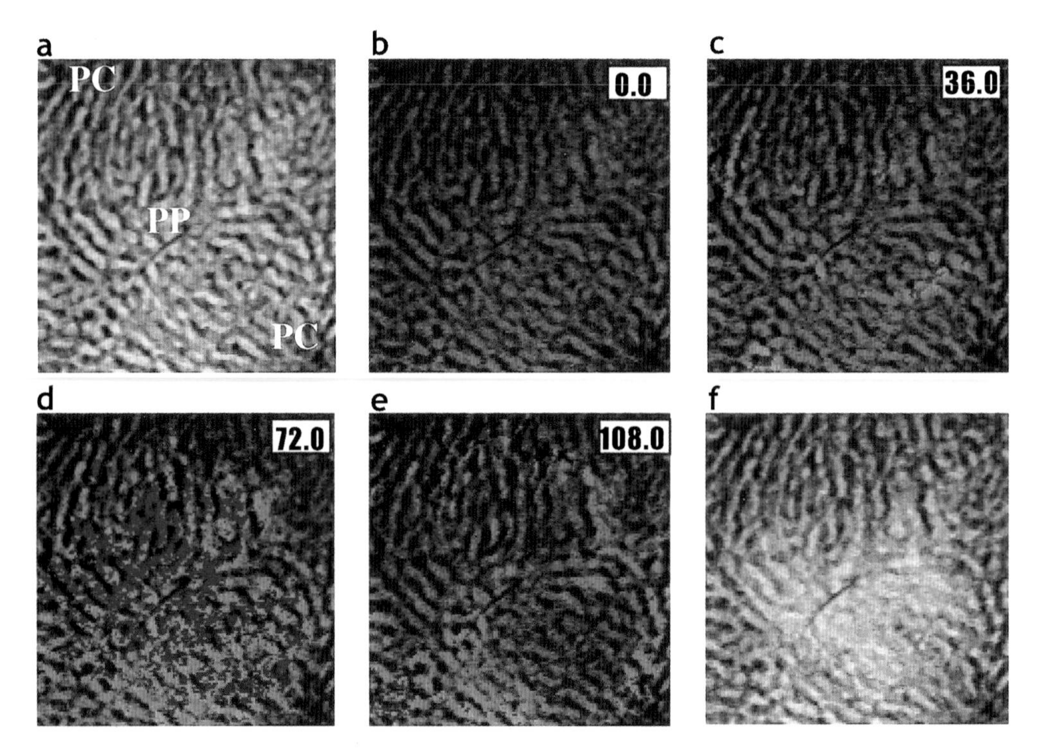

Plate 9 Propagation of a coordinated intercellular $[Ca^{2+}]_i$ wave in the intact liver. Confocal images of fluo-3 fluorescence were acquired prior to and after infusion of an intact perfused rat liver with 80 pM vasopressin. (a) The initial fluo-3 fluorescence intensity depicted as a linear grey scale. The periportal (PP) and pericentral (PC) regions of the lobule are marked. (b–e) Images show the fluo-3 fluorescence intensity changes normalized to the initial intensity (i.e. F/F_0) displayed on a pseudocolour scale (a change from blue to red represents a ratio scale range of 0.8 to 1.45). The numbers in the upper right-hand corner represent time, in seconds, after vasopressin stimulation. (f) Grey scale fluo-3 fluorescence image overlaid with a red difference image showing the initial sinusoidal staining with fluorescein-labelled bovine serum albumin (F-BSA). The difference images were calculated by subtracting sequential fluorescence images through the time series. The intensity of the overlay is proportional to the magnitude of the fluorescence change at each pixel. The distribution pattern of F-BSA is used to determine the PP and PC zones of the lobule. The images were obtained using a Zeiss Plan-NEOFLUAR × 16 objective and a 40 μm optical section. The image size is 388 × 388 μm.

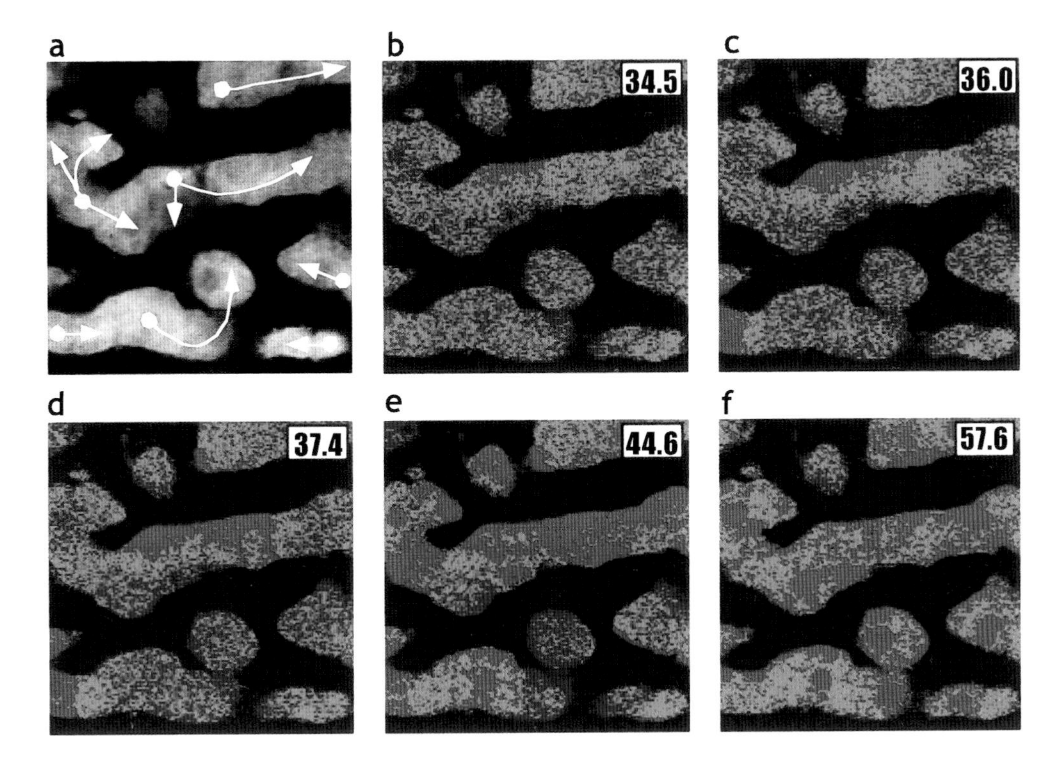

Plate 10 IP_3-dependent intercellular $[Ca^{2+}]_i$ waves propagate along hepatic plates. (a) The initial frame is a confocal image of the fluo-3 fluorescence intensity shown as a linear grey scale. The white dots show the initiating hepatocyte and arrows show the direction of a intercellular $[Ca^{2+}]_i$ wave. It should be noted that intercellular $[Ca^{2+}]_i$ waves may well originate from outside of the focal plane, due to the three-dimensional structure of the intact tissue, and thus the actual propagation pathway may not be accurately described in terms of $\times$ and y coordinates. (b–f) Pseudocolour images illustrating intracellular and intercellular $[Ca^{2+}]_i$ waves propagating along the hepatic plate during stimulation with 160 pM vasopressin. The fluo-3 fluorescence intensity changes are displayed as an F/F_0 ratio image as described in Plate 9 (see page 000). The numbers in the upper right-hand corner represent time, in seconds, after vasopressin stimulation. The images were obtained using a Olympus SPlan Apo $\times$ 60 objective and a 1.0 μm optical section. The image size is 104 $\times$ 104 μm.

Table 1 Comparison of optical detection methods for single cell gene expression

Reporter	Substrate	Advantages	Disadvantages
β-galactosidase	X-gal, FDG	Enzymatic amplification, FACS-compatible Standard fluorescein optics	Substrate not membrane-permeant, poorly retained
β-lactamase	CCF-2	Enzymatic amplification, FACS-compatible Membrane-permeant substrate Ratiometic read-out	Requires violet excitation, for optimal performance
Luciferase	Luciferin	Large dynamic range	Relatively few emitted photons/time
		No optical background	Requires longer read-out times Not FACS-compatible
GFP	None	Requires no substrate Several colours available	Non-enzymatic, therefore relatively insensitive

genetic construct. There are numerous reporter systems including β-galactosidase, choline acetyl transferase, secreted alkaline phosphatase, luciferase, and β-lactamase (see *Table 1*). However, only luciferase, β-galactosidase, β-lactamase have been used at the single cell level. Single cell detection has several requirements. First, the reporter enzyme and its substrate must be introduced into, and retained by, the cell long enough to obtain a measurement. Secondly, the enzyme–substrate system should have appropriate kinetic features to ensure a reliable measurement in cells. For example, because the concentration of product could be high relative to substrate in cells expressing the reporter, it is preferred that the product not inhibit the enzyme. Finally, the read-out should be compatible with commonly used detection instrumentation. Optical read-outs such as fluorescence are preferred due to the relatively bright signal and the compatibility with fluorescence activated cell sorting (FACS).

2.1.1 β-galactosidase

Historically, β-galactosidase was the first enzyme to be used as a single cell reporter. Two assays allow for the detection of β-galactosidase activity in single cells: X-gal and fluorescein-β-di-galactopyranoside (FDG). X-gal is a colorimetric substrate that yields a blue precipitate when cleaved by β-galactosidase and is usually used to examine fixed tissues. Because the protocol is difficult to carry out and quantify using digital imaging microscopes or fluorescence activated cell sorters, this approach is generally not used for measuring gene induction in response to Ca^{2+}. An alternative substrate is FDG which yields free fluorescein upon cleavage by β-galactosidase. The fluorescent product of FDG can be detected in live cells by flow cytometry and by epifluorescence microscopy. FDG has been widely used in flow cytometry but has several shortcomings as a means of monitoring gene expression over time in single cells. First, it is highly polar, and therefore does not cross cell membranes easily. Consequently it must be loaded by osmotically shocking cells which is inefficient and causes inconsistent loading between cells. This is a concern because the observed fluorescence intensity is a function both of the amount of β-galactosidase in the cell and of the amount of substrate; therefore differences in fluorescence are difficult to interpret. Secondly, the cleavage product, fluorescein, only has one charge and leaks rapidly out of cells at room temperature. Thus, cells must be incubated on ice to permit the product to accumulate without leaking. Several attempts have been made to make FDG a better substrate for room temperature measurements in intact cells including adding extra carbons to make the product more lipophilic or adding functional groups that will allow the product to bind intracellular proteins (these are available from Molecular Probes). These modifications have their own problems including slow cleavage by the enzyme and poor fluorescence. Currently the best method for analysing gene expression in single cells uses β-lactamase (described below), however β-galactosidase may be useful under some circumstances so the protocol is provided.

Protocol 1

Measuring gene expression using FDG on a microscope

Equipment and reagents

- Microscope
- 2 mM FDG
- PBS

Method

1 Prepare 2 mM FDG solution by dissolving in H_2O and heating at 30 °C until dissolved. Store frozen in aliquots.

2 For experiments on a microscope, introduce a loading solution of 50/50 FDG stock and PBS to the cells for 30–45 sec. Conclude the reaction using cold, slightly hypotonic (80% tonicity) PBS solution to prevent damage due to excessive cell shrinkage upon conclusion of the hypotonic shock. If possible, incubate cold for 10–15 min, otherwise incubate at room temperature for 2–5 min.

3 Image cells using excitation filters and dichroic mirrors for fluorescein (ex. 475 nm; DC 490; em. 500 LP).

Protocol 2

Loading cells with FDG for flow cytometer measurements

Equipment and reagents

- Microcentrifuge tubes
- 2 mM FDG
- PBS
- Propidium iodide

Method

1 Place 50 μl of 5×10^6/ml cell in a microcentrifuge tube and dilute with an equal volume of the 2 mM FDG stock in H_2O.

Protocol 2 continued

2. Incubate for 30 sec at 37 °C (the timing is critically important for obtaining reproducible loading).
3. End the reaction by adding 450 μl of ice-cold, isotonic PBS.
4. Incubate cells on ice for 80 min.
5. For FACS analysis, concentrate the cells by centrifugation, resuspending in 100 μl of cold PBS, and adding propidium iodide to identify dead cells.

2.1.2 β-lactamase

A reporter system based on the bacterial antibiotic resistance gene, β-lactamase is a convenient alternative to the β-galactosidase system because it employs a membrane-permeant, cytoplasmic-trapped fluorogenic substrate called CCF-2 (available from Aurora Biosciences Corp.) (6). The design of the substrate utilizes fluorescence resonance energy transfer (FRET) to produce a ratiometric read-out. The substrate is composed of a 7-hydroxycoumarin linked to a fluorescein by a cephalosporin core. The substrate is chemically modified such that ester groups on the molecule are masked by butyrate, acetate, and acetoxymethyl ester groups to make the molecule both lipophilic and non-fluorescent. The substrate readily crosses cell membranes. Inside the cell, hydrolysis and enzymatic processing of the substrate generate a trapped fluorescent molecule. As a result of this accumulation mechanism, the extracellular CCF-2 concentration can be in the μM range (approximately 1000-fold lower than the concentration of FDG required to load cells). In the intact molecule, excitation of the coumarin results in efficient FRET to the fluorescein, resulting in green fluorescence. Cleavage of CCF2 by β-lactamase results in spatial separation of the two dyes, disrupting FRET, and causing cells to change from green to blue when viewed using a fluorescence microscope.

The cellular retention of both substrate and product and the lack of any enzyme inhibition by the product allow the blue signal to develop over time and yields a detection limit of about 50 enzyme molecules/cell. This permits the detection of genes with low expression levels in single cells. In addition, it has been possible to load unstimulated cells with substrate and observe activation of gene expression at the single cell level in real time (6). The compatibility with fluorescence activated cell sorting (FACS) is quite useful because it permits the efficient isolation of responsive clones from heterogeneous cell pools. The ratiometric read-out offers advantages over simple intensity-based assays. Since the read-out is a colour change, the ability to resolve positive and negative events is greatly improved due to the fact the low expressing cells are as fluorescent as high expressors, and differ only in the wavelength of their emission. The system is therefore well suited for measuring both induction and repression of gene expression. Recently, a promoterless β-lactamase vector has been introduced into cell lines where it has integrated downstream of endogenous response elements (7). Using iterative enrichment by FACS, clones responsive to various signals, including Ca^{2+} can be isolated. This technique will prove useful in producing reporter lines for endogenous genes and helping to further explore the ways in which different genes are regulated by Ca^{2+}.

Protocol 3

Loading cells with CCF-2/AM

Reagents

- CCF-2/AM (Aurora Biosciences)
- Dry DMSO
- Pluronic-F127® (Molecular Probes)
- Acetic acid
- Serum-free medium containing Hepes (25 mM, pH 7.3), e.g. RPMI (Irvine Scientific) or HBS (pH 7.3)

Method

1. Dissolve CCF-2/AM to 1 mM in DMSO. This stock solution can be frozen and should be protected from light for future use (for about one month).
2. Prepare 1 ml of dispersant solution containing 100 mg of the dispersant Pluronic-F127®, 100 μl acetic acid (0.1%), and 900 μl DMSO.
3. Mix 1 μl of CCF-2/AM stock solution with 9 μl of the dispersant solution.[a]
4. Add the resulting solution with vigorous agitation to 1 ml of serum-free medium containing Hepes (25 mM, pH 7.3) such as RPMI or HBS (pH 7.3).
5. Add the resulting medium to cells for 1–2 h at room temperature.

[a] CCF-2/AM has limited stability in aqueous solutions. Therefore the loading buffer should be used within 1 h of its preparation.

2.1.3 GFP

Recently, bright mutants of green fluorescent protein (GFP) have been employed as reporter genes (8). The main advantage of GFP as a reporter is that no substrate is required for detection. In addition, there are several spectrally resolved GFPs, which allows the tracking more than one gene in the same cell (9). However, GFP is non-enzymatic and thus has a relatively low sensitivity. The limit of GFP detection is about 1×10^5 molecules/cell (8) primarily because of the autofluorescent background of the cell. Although examples of GFP-based reporters are emerging, the low sensitivity constrains the use of GFP as a reporter for many low-expressing endogenous promoters.

2.1.4 Luciferase

Luciferase acts on a luminescent substrate, luciferin, to produce light. Although the amount of light produced by a luminescent substrate is several orders of magnitude lower than the amount of light produced by a fluorescent molecule, the signal-to-noise is comparable to a fluorescence assay due to a very low background. Although luciferase is usually assayed in cell supernatants, luciferin can be loaded into cells using high extracellular concentrations and can therefore be used to detect gene expression in single cells (10). However, rather long integration times using cooled CCDs are required to quantify the signal and the method is incompatible with either FACS or visual observation. As a result, luciferase is unlikely to be widely used as a readout for gene expression in single cells.

3 Generating defined cytoplasmic $[Ca^{2+}]_i$ signals

The overall goal of the Ca^{2+} clamp is to generate a temporally defined $[Ca^{2+}]_i$ pattern in a large population of cells. This is achieved by making the cell membranes highly permeable to Ca^{2+} and then changing the concentration of extracellular calcium ($[Ca^{2+}]_o$). Changing the amount of Ca^{2+} in the media makes it possible to modulate the $[Ca^{2+}]_i$ concentration generating a signalling pattern that changes with time. In Ca^{2+}-permeabilized cells, high $[Ca^{2+}]_o$ leads to a large Ca^{2+} influx across the membrane that exceeds the rate of Ca^{2+} efflux and produces a cytoplasmic calcium rise. Conversely, when $[Ca^{2+}]_o$ is low then the rate of influx is reduced and Ca^{2+} clearance from the cell causes a drop in $[Ca^{2+}]_i$. Because the solution surrounding a large number of cells can be changed almost synchronously and because many cells can be made permeable to Ca^{2+}, it is possible to generate patterns in large numbers of cells at once. In addition, because cells can be made permeable to Ca^{2+} in ways that do not activate other signalling pathways, it is possible to analyse the effects of Ca^{2+} signalling largely independently of other second messengers. Below we describe the theory, construction, and testing of a Ca^{2+} clamp that we have used (11) to study Ca^{2+} signalling in T lymphocytes and modifications that will make it useful for other cell types.

3.1 Constructing a calcium clamp

A calcium clamp experiment has three main steps:

(a) Making cells permeable to extracellular Ca^{2+}.

(b) Perfusing a large number of cells uniformly.

(c) Measuring $[Ca^{2+}]_i$ with time-lapse video microscopy.

These three steps are discussed in detail in the following sections.

3.1.1 Calcium permeabilization of cells

There are two general strategies for making cells selectively permeable to calcium: using Ca^{2+} ionophores and using endogenous Ca^{2+} channels. The first relies on the use of compounds like A23187 and ionomycin which transport Ca^{2+} in and out of cells. At high concentrations ($> 1\ \mu M$) ionophores are very effective at mobilizing Ca^{2+} from internal stores and at transporting Ca^{2+} across cell membranes. Ionophores are useful for experiments in which the extracellular solution is exchanged only a few times. However for long experiments the cost becomes prohibitive because they must be included continuously in the extracellular media which may be exchanged hundreds of times throughout an experiment. A further difficulty is that ionophores are unstable at 37 °C so that over time they transport Ca^{2+} less efficiently making it difficult to make predictable changes in $[Ca^{2+}]_i$ late in an experiment.

A second approach relies on the use of Ca^{2+} channels which can be activated by voltage or pharmacological agents. The main advantage of endogenous channels is that they are the major source of sustained $[Ca^{2+}]_i$ signals in many cells and therefore recreate the local $[Ca^{2+}]_i$ gradients which occur following physiological activation. In addition, there are irreversible and inexpensive ways of activating endogenous Ca^{2+} channels that have minimal effects on other signalling pathways.

In our experiments we used store-operated Ca^{2+} channels to make the cell membrane of lymphocytes permeable to Ca^{2+}. Store-operated channels are the predominant Ca^{2+} influx pathway in non-excitable cells. These channels are activated irreversibly by thapsigargin, which inhibits the SERCA family of Ca^{2+} ATPases and causes depletion of the internal stores. We treated T lymphocytes with 1 μM thapsigargin for 5 min in low calcium RPMI (see *Protocol 4*). This treatment was sufficient to fully activate the store-operated Ca^{2+} channels for the three hour duration of our experiments. Possible disadvantages of using thapsigargin are that it has been reported to activate stress-activated signalling cascades in some cell types and that it may inhibit the translation of some proteins by reducing endoplasmic reticulum Ca^{2+} levels. This does not appear to be a major problem in T lymphocytes in which robust protein expression occurs even in the presence of maximal concentrations of thapsigargin.

In excitable cells, such as neurons and myocytes, voltage-gated Ca^{2+} channels or glutamate receptors could be used to permeabilize the membrane to Ca^{2+}. There are many ways of activating these channels including treating with NMDA and depolarizing with potassium. It is important to note that these channels are not uniformly distributed throughout the cell membrane and may generate local $[Ca^{2+}]_i$ gradients which need to be taken into account when interpreting the experimental results. Finally, if suitable endogenous Ca^{2+} channels are not available, it may also be possible to transfect cells with exogenous channels that can be used to make the membrane permeable to Ca^{2+}.

3.1.2 Perfusing cells uniformly

The second component of the Ca^{2+} clamp is a perfusion system that can rapidly and uniformly change the $[Ca^{2+}]_o$ surrounding a large population of cells. Rapid perfusion is necessary to generate the high frequency $[Ca^{2+}]_i$ spikes that are often observed in stimulated cells, while uniform perfusion is essential to ensure that the $[Ca^{2+}]_i$ concentrations are constant across a field of cells. The importance of uniformity can not be overstated as even small heterogeneities in the solution exchange can lead to dramatic differences in the peak $[Ca^{2+}]_i$ levels achieved in different cells. The perfusion system can be divided into the mixing system that delivers heated and pH buffered solutions containing different Ca^{2+} concentrations to the cells, and the laminar flow chamber which allows the rapid and uniform perfusion of the cells.

i. Mixing system

The mixing system (*Figure 1A*) is composed of two 500 ml glass bottles placed in a heated water-bath (set slightly above 37 °C to allow for cooling of the solution in transit to the cells) and pressurized with 5% CO_2 and 95% air to 2 psi. One bottle contains a high calcium solution (see *Protocol 4*) and a second bottle contains an identical solution that lacks the added calcium and contains 0.5 mM EGTA. Each bottle is capped with a two-hole rubber stopper that contains an input from a regulated source of air and CO_2 and a flexible polypropylene straw (o.d. 1 cm, i.d. 0.5 cm) that reaches the bottom corner of the bottle and extends above the rubber stopper about an inch. Each polypropylene tube is connected via silicone rubber tubing to a regulated flow valve (Regulator IV extension set 1671, Abbot Laboratories) that is connected to the input ports of a three-way solenoid valve (valve 3-132-900 with driver 90-1-100; General Valve Corp.). The flow valves are necessary to make fine adjustments to the flow of solution into the chamber and the three-way solenoid valve determines which of the two solutions flows over the cells. One of the lines from the flow valve to the solenoid valve is interrupted by a three-way Luer stopcock (Catalog No 30600-04, Cole Parmer Instruments) that serves as an injection port. The output of the solenoid valve is connected by a 1 cm tube (o.d. 2.4, i.d. 0.8) to a Y junction which has two outputs that lead directly into a laminar flow chamber.

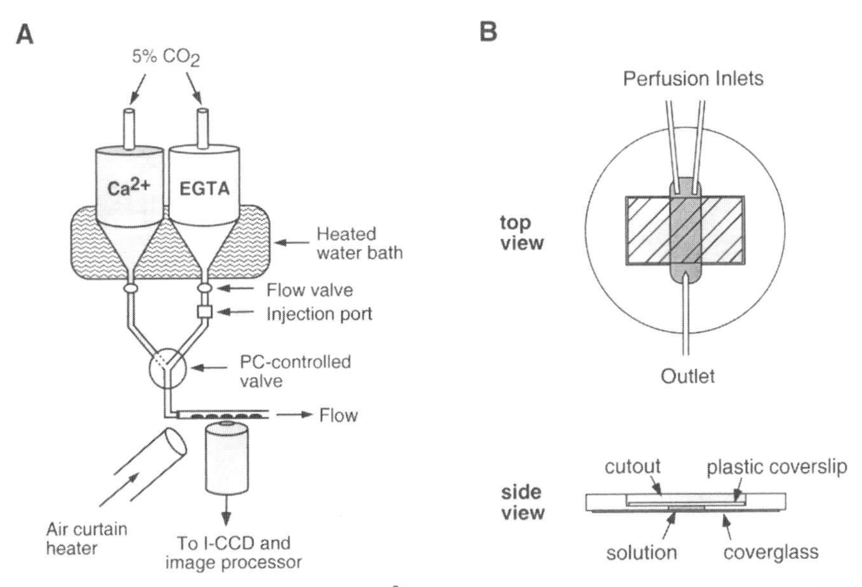

Figure 1 (A) Schematic diagram of the Ca^{2+} clamp. Solution flows from the two reservoirs through the PC-controlled Y valve and over the cells on the laminar flow chamber. (B) Top and side views of the perfusion chamber. Taken with permission from ref. 11.

During an experiment, Ca^{2+} flows from the bottles through the flow valves and into the computer-controlled solenoid valve. By controlling the valve, the computer can direct which of the two solutions flows into the flow chamber perfusing the cells. In the simplest case, it is possible to generate periodic $[Ca^{2+}]_i$ increases by intermittently perfusing Ca^{2+} permeabilized cells with high $[Ca^{2+}]_o$. However, it is also possible to generate intermediate $[Ca^{2+}]_o$ by rapidly switching between the two solutions and allowing them to mix in the tubing that connects the valve to the chamber. The longer the valve spends in the high Ca^{2+} position the higher the concentration of Ca^{2+} in the solution that flows into the chamber. In this way, the computer can control the exact $[Ca^{2+}]_o$ that is flowing over the cells.

ii. Laminar flow chamber

The laminar flow chamber (*Figure 1B*) consists of a No 1 glass coverslip permanently mounted with Sylgard (Dow Corning) across a rectangular opening (6 $\times$ 20 mm) in a Lucite disk. A plastic coverslip is mounted using silicone grease on the opposite side of the Lucite disk. The two coverslips are separated by a thin lip of plastic as well as by the grease, creating a flow chamber with dimensions 13 mm $\times$ 6 mm $\times$ 0.8 mm (total volume 60 μl). At one end of the chamber, two stainless steel 18 gauge needles flow directly into the space between the two coverslips while at the other a blunt needle is connected to a suction line.

The chamber described above is easily manufactured in a basic machining shop. We have confirmed that the flow through this chamber is essentially laminar and uniform by measuring the rate of solution exchange using fluorescein and by observing the exchange patterns using ruthenium red (see Testing and troubleshooting the Ca^{2+} clamp, below). At a flow rate of 3.5 ml/min 99 $\pm$ 5% exchange is achieved within 3 sec. Alternatively, laminar flow chambers that are suitable for this purpose can be purchased from Warner Instruments.

3.1.3 Time lapse imaging of calcium

The third component of the calcium clamp is a Ca^{2+} imaging system which allows the investigator to verify that cells are responding in a uniform fashion to changes in $[Ca^{2+}]_o$. We have mounted the laminar flow chamber on a standard ratiometric Ca^{2+} imaging microscope and we have loaded the cells with a fura-2 derivative.

i. Imaging system

Since the methods for measuring $[Ca^{2+}]_i$ in single cells have been described extensively in several other publications (and earlier chapters of this book) we will describe them here only briefly. Our system (12) consists of an inverted microscope (Zeiss Axiovert 35) equipped for epifluorescence, connected to an intensified CCD camera (Hamamatsu Corp.) and an image processor (Video Probe, ETM Systems). A computerized filter wheel (Lamda-10, Sutter Instruments) alternates between two excitation wavelengths (350 and 380), the camera collects the two images, and the image processor corrects for background, divides the two intensity images on a pixel-by-pixel basis, and calculates the cytoplasmic calcium concentration which is displayed in pseudocolour. A series of ratio images are collected during an experiment (e.g. every 5 sec). The computer also controls the opening and closing of the solenoid valve via TTL logic pulses delivered to a valve control unit. Any of a number of commercially available ratio imaging systems can be used to construct a Ca^{2+} clamp. The main requirements are that images be captured and processed in close to real time to provide a diagnostic feedback during an experiment and that the system be capable of delivering TTL pulses. Our system is also equipped with a home-made air curtain heater which is essentially a thermostatically controlled hair dryer that blows air under the stage of the microscope and keeps the objectives and the flow chamber at 37 °C.

ii. Calcium dyes

There are a number of Ca^{2+}-sensitive dyes that can be used for $[Ca^{2+}]_i$ imaging and their various advantages and disadvantages have been described previously (13). We initially chose fura-2 because it is the simplest and most quantitative way of measuring $[Ca^{2+}]_i$ using an imaging microscope. However, we found that fura-2 leaks rapidly from T cells at 37 °C so that almost all the signal is lost after one hour. This problem can be overcome by using fura-PE3 (TefLabs) a derivative of fura-2 that is more resistant to export or leakage from cells. The AM ester of fura-PE3, however, is more difficult to load into lymphocytes and we found it necessary to increase both the concentration and incubation time as well as to reduce the fetal calf serum in the loading medium (see *Protocol* 4). We routinely record $[Ca^{2+}]_i$ in fura-PE3 loaded cells for up to 3 h at 37 °C.

3.2 Using the calcium clamp

In the section below, we provide protocols for a generic calcium clamp experiment done using T lymphocytes and the system described above.

Protocol 4

Running a generic calcium clamp experiment

Equipment and reagents

- See *Protocols* 5–9

Method

1. Load the cells with fura-PE3/AM (see *Protocol* 5).
2. Coat the glass coverslip on the laminar flow chamber with poly-D-lysine (see *Protocol* 6).
3. Prepare and degas the high calcium and low calcium experimental solutions (see *Protocol* 7). Degassing the solutions is necessary to prevent bubbles from forming in the solution and washing the cells off the chamber during the experiment.
4. Initialize and prime the perfusion system (see *Protocol* 8).

Protocol 4 continued

5 Place the cells on the poly-D-lysine coated coverslip, assemble the chamber, and connect it to the perfusion system (see *Protocol 9*).

6 Program the computer to open and close the solenoid valve at the appropriate times to deliver the correct calcium stimulus.

7 Initialize the imaging system by collecting a background and selecting a field of cells to image.

8 Start collecting ratio images.

9 Calcium permeabilize cells by injecting low calcium RPMI solution containing 1 μM thapsigargin via the Luer stopcock (injection port). Monitor the release of Ca^{2+} from the internal stores.

10 When the release is complete (after about 5 min), start the constant perfusion of the cells. This can be done either by manually turning the Luer stopcock to allow the pressurized liquid to flow over the cells or by programming the computer so that the solenoid valve is switched at the appropriate moment. Solution flow should not start before the cells have been treated with thapsigargin for at least a minute as this will wash the inhibitor from the chamber prematurely.

11 Monitor $[Ca^{2+}]_i$, changing the imaging field periodically to verify that perfusion is occurring uniformly across the chamber.

12 Turn off the perfusion system and image gene expression (see Section 2) or wash the cells off the chamber and analyse them biochemically.

13 Analyse the $[Ca^{2+}]_i$ imaging data off-line.

14 Wash the perfusion system by replacing both the low and high calcium solutions with distilled water. Wash at least 25 ml of water through the system.

Protocol 5

Loading cells with fura-PE3/AM

Equipment and reagents

- Microcentrifuge
- Loading buffer: RPMI 1640, 25 mM Hepes, 2% fetal calf serum
- Fura-PE3/AM stock solution: 1 mM in DMSO

Method

1 Prepare fura loading solution by adding 2 μl of fura-PE3/AM to 1 ml of loading buffer and vortexing for 30 sec.

2 Spin down 10^6 cells for 3 sec in a microcentrifuge at maximum speed, remove the supernatant, and resuspend in fura loading solution.

3 Incubate at 37 °C for 1 h with gentle mixing every 15 min.

Protocol 5 continued

4 Wash the cells three times with 1 ml of loading solution (without fura-PE3) before resuspending in 100 μl of loading solution.

5 Incubate the cells for 1 h at 37 °C to allow full de-esterification of the dye.

Protocol 6

Coating the laminar flow chamber with poly-D-lysine (PDL)

Equipment and reagents

- Coverslip
- Ethanol
- PDL solution (poly-D-lysine): 1 mg/ml in dH_2O

Method

1 Carefully wash the glass coverslip on the chamber with ethanol and then with double distilled H_2O scrubbing gently with a cotton tipped swab.

2 Distribute 60 μl of PDL solution evenly across the glass and incubate for 15 min in a humidified environment.

3 Aspirate off PDL solution and wash once with distilled H_2O before aspirating dry.

Protocol 7

Preparing and degassing perfusion solutions

Equipment and reagents

- 500 ml tissue culture glass bottles
- Shaking water-bath
- RPMI 1640 without phenol red, biotin, or riboflavin (as these contribute to autofluorescence; Irvine Scientific)
- 1 M $CaCl_2$
- 1 M Hepes pH 7.4
- 1 M EGTA
- 5% bovine serum albumin (Sigma Chemical Tissue Culture Grade)
- High calcium RPMI: RPMI, 25 mM Hepes, 0.1% BSA, 1 mM $CaCl_2$ (1.5 mM calculated total Ca^{2+})
- Low calcium RPMI: RPMI, 25 mM Hepes, 0.1% BSA, 0.5 mM EGTA (50 μM calculated total Ca^{2+})

Method

1 Make the appropriate volume of high calcium RPMI (HCR) and low calcium RPMI (LCR) in 500 ml tissue culture glass bottles. Calculate the volume of the solution required for the experiment by multiplying the length of the experiment by the flow rate of the solution and adding 10%.

2 Close the bottles with one-hole rubber stoppers in which the hole is covered by a three-way Luer stopcock.

Protocol 7 continued

3 Connect a vacuum line to the stopcock, swirl the solution in the bottle, and close the stopcock, capturing the vacuum in the bottle.
4 Place the vacuum sealed bottles in a 37 °C shaking water-bath for 1 h before the start of the experiment.

Protocol 8

Priming the perfusion system

Equipment and reagents

- Perfusion system
- See *Protocol 7*

Method

1 Place the two-hole rubber stoppers with the plastic straws into the bottles containing the high and low Ca^{2+} solutions.
2 Place a 3.5 cm tissue culture dish on the microscope and place the two efflux tubes that are downstream of the solenoid valve above the dish.
3 Connect the bottles to the source of CO_2/air and pressurize the bottles to 2 psi.
4 Fill the perfusion tubing with solution by allowing the solution to flow through the system until it drips out of the efflux tubes. Switch the solenoid valve to allow filling of both perfusion paths and to remove air bubbles trapped in the valve.
5 Remove any air bubbles that are trapped in the tubing by tapping gently and by repeatedly switching the solenoid valve.
6 Adjust the solution flow rate to approx. 3.5 ml/min by adjusting the flow valves. Measure the flow rate with the aid of a timer and a small graduated cylinder.
7 Stop the flow by closing the valve on the injection port and switching the solenoid valve.

Protocol 9

Assembling the perfusion chamber

Equipment and reagents

- Syringe and needle
- Poly-D-lysine coated glass coverslip
- Silicone grease
- Fura-PE3 loaded cells

Method

1 Add 100 μl of fura-PE3 loaded cells to the poly-D-lysine coated glass coverslip, taking care to distribute them evenly but avoiding the extreme corners which are poorly perfused. Allow the cells to settle for 10 min.

Protocol 9 continued

2. Load a 5 ml syringe from the back with silicone grease. Attach a blunt 20 gauge needle to the syringe and carefully deposit a thin bead of silicone grease on the lip that separates the lower coverslip from the upper coverslip.
3. Press the top coverslip onto the silicone grease to form firm seal between the coverslip and the chamber.
4. Mount the assembled chamber on the microscope stage and adjust the influx tubes to point to the region between the two coverslips.
5. Place the needle attached to the vacuum line so it will suction solution flowing through the chamber but will not dry the cells.

3.3 Characterizing and troubleshooting the Ca^{2+} clamp

The performance of the Ca^{2+} clamp depends critically on the speed and uniformity of solution exchange in the chamber. There are three ways of testing the solution exchange. One method is to add 1 mM phenol red to one of the perfusion solutions and to examine the pattern of solution exchange in the laminar flow chamber shortly after switching the solenoid valve. Because the red dye is easily detected by eye, this method can be used to examine whether all the regions of the chamber are exchanged uniformly. A second approach is to add a diluted solution of fluorescein to one of the perfusion bottles and to use the fluorescence imaging system to quantify the efficiency of exchange in different regions of the chamber. It is good practice to scan various regions of the chamber and to calculate and compare the rate of exchange periodically during the life of the calcium clamp to verify that exchange is occurring efficiently. A final way of examining the exchange rate is to measure the rate and amplitude of $[Ca^{2+}]_i$ increase in fura-2 loaded, calcium permeabilized cells following 50 second pulses of high $[Ca^{2+}]_o$. Because the cells are sensitive indicators of $[Ca^{2+}]_o$, changes in the peak height of the resulting spikes provide an effective means of determining differences in $[Ca^{2+}]_o$ in different regions of the chamber.

In addition to monitoring the efficacy of perfusion it is also important to verify that the temperature in the perfusion chamber remains relatively constant throughout an experiment. The easiest way of doing this is to use a miniature thermocouple (Omega Corp.) connected to a digital thermometer. The thermocouple can be inserted between the coverslips of the perfusion chamber and used to monitor the temperature in the chamber during the experiment. The temperature in the chamber is determined both by the temperature of the solutions in the water-bath and by the temperature of the air curtain heater which heats the microscope. Because the temperature of the solution falls from the time it leaves the bottles to the time it enters the chamber it is necessary to heat the water-bath to a temperature slightly higher than 37 °C. It is important to verify that the temperature of the solution coming from each of the bottles does not vary by more than a few degrees. This can occur if one of the solutions spends more time out of the bottle and in transit to the chamber than the other because of differences in flow rate, or because the computer controlled valve spends more time in one of the two positions. In this case, it may be necessary to insulate the tubing leading up to the valve with neoprene or styrofoam. With some adjustment, it is possible to prevent major fluctuations in temperature resulting from switching of the solenoid valve.

3.4 Summary and future perspectives

The calcium clamp described above is a powerful technique to generate relatively uniform and synchronous $[Ca^{2+}]_i$ patterns in a population of cells. It relies on changing Ca^{2+} influx through channels in the cell membrane and in this way may reconstitute naturally-occurring $[Ca^{2+}]_i$ gradients in cells in which the majority of the $[Ca^{2+}]_i$ rise results from influx across the membrane. The ability to generate defined, relatively uniform Ca^{2+} signalling patterns coupled to methods for measuring gene expression in single cells, should be useful in unravelling the Ca^{2+} signalling pathways in a variety of cells.

Variations on the methods described here include using changes in membrane potential to trigger Ca changes, through either voltage-gated or store-operated Ca influx mechanisms. Enhancements to these methods include using controlled feedback between the imaging system and the calcium clamp fluidics to control the $[Ca^{2+}]_i$ pattern more precisely. These modifications will increase the versatility and the precision of the calcium clamp and its applicability to the study of time- and amplitude-dependent responses in gene expression.

References

1. Ghosh, A. and Greenberg, M. (1995). *Science*, **268**, 239.
2. Berridge, M. J. (1993). *Nature*, **361**, 315.
3. Tsien, R. W. and Tsien, R. Y. (1990). *Annu. Rev. Cell Biol.*, **6**, 715.
4. Fewtrell, C. (1993). *Annu. Rev. Physiol.*, **55**, 427.
5. Bronstein, I., Fortin, J., Stanley, P., Stewart, G., and Kricka, L. (1994). *Anal. Biochem.*, **219**, 69.
6. Zlokarnik, G., Ñegulescu, P., Knapp, T., Mere, L., Burres, N., Feng, L., *et al.* (1998). *Science*, **279**, 84.
7. Whitney, M., Rockenstein, E., Cantin, G., Knapp, T., Zlokarnik, G., Sanders, P., *et al.* (1998). *Nature Biotechnol.*, **16**, 1329.
8. Hiem, R. and Tsien, R. (1996). *Curr. Biol.*, **6**, 178.
9. Yang, T., Sinai, P., Green, G., Kitts, P. A., Chen, Y., Lybarger, L., *et al.* (1998). *J. Biol. Chem.*, **273**, 8212.
10. Rutter, G., White, M., and Tavare, J. (1995). *Curr. Biol.*, **5**, 890.
11. Dolmetsch, R. E. and Lewis, R. S. (1999). In *Imaging living cells: a laboratory manual* (ed. A. Konnerth, F. Lanni, and R. Yuste). CSHL Press, Cold Spring Harbor, p. 279.
12. Dolmetsch, R., Xu, K., and Lewis, R. (1998). *Nature*, **392**, 933.
13. Grynkiewicz, G., Poenie, M., and Tsien, R. Y. (1985). *J. Biol. Chem.*, **260**, 3440.

Chapter 9

Monitoring generation and propagation of calcium signals in multicellular preparations

Lawrence D. Gaspers and Andrew P. Thomas
Department of Pharmacology and Physiology, University of Medicine and Dentistry of New Jersey, Newark, New Jersey 07103, USA.

Sandip Patel
Department of Pharmacology, University of Oxford, Oxford OX1 3QT, UK.

Anthony J. Morgan
Centre for Cardiovascular Biology and Medicine, GKT School of Biomedical Sciences, King's College London, London SE1 1UL, UK.

Paul A. Anderson
Department of Pathology, Anatomy and Cell Biology, Thomas Jefferson University, Philadelphia, PA 19017, USA.

1 Introduction

That this book is devoted to the measurement of intracellular calcium ions (Ca^{2+}) underscores the pivotal role this cation plays in regulating a diverse array of cellular events from fertilization to apoptosis (1). Indeed, Ca^{2+} may be responsible for your first visit to the obstetrician and your last visit to the pathologist. Transient spikes or oscillations in cytosolic free calcium concentration ($[Ca^{2+}]_i$) are utilized by many different extracellular stimuli to activate or modify intracellular Ca^{2+}-sensitive processes, which ultimately lead to changes in tissue function. Ca^{2+} exerts its effects by either directly binding to Ca^{2+}-sensitive target proteins or indirectly through Ca^{2+}-binding proteins such as calmodulin (2). A common mechanism by which extracellular agonists alter $[Ca^{2+}]_i$ homeostasis is through activation of inositol lipid signalling cascade and subsequent production of the second messenger, inositol 1,4,5-trisphosphate (IP_3). IP_3 released into the cytosol binds to IP_3 receptors (IP_3R) to mobilize internal Ca^{2+} stores (3–5), that in turn stimulates the influx of external Ca^{2+} to sustain the signal and refill the internal stores (6).

In non-excitable cells, Cobbold and co-workers (7, 8) were the first group to directly demonstrate that activation of hormone receptors coupled to phospholipase C generate periodic baseline-separated $[Ca^{2+}]_i$ oscillations in aequorin-injected hepatocytes. This was subsequently confirmed in our laboratory using digital imaging techniques (9). At the single cell level, IP_3-dependent Ca^{2+} signals show remarkable complexity; each $[Ca^{2+}]_i$ spike arises from a specific subcellular locus and propagates in a regenerative manner throughout the cytoplasm as an intracellular $[Ca^{2+}]_i$ wave (10). Although the regulatory mechanisms that give rise to $[Ca^{2+}]_i$ spikes and $[Ca^{2+}]_i$ waves are still being actively investigated, these phenomena are thought to be elicited by periodic IP_3- and Ca^{2+}-mediated activation and inactivation of the IP_3R (3–5, 11).

The study of the complex spatial and temporal organization of Ca^{2+} signals has been greatly facilitated by the synthesis of fluorescent Ca^{2+}-sensitive indicator dyes and development of imaging techniques capable of measuring small fluorescence changes at the cellular and subcellular level (see Chapter 2 in this volume). The vast majority of work has been conducted in single isolated cells or cell monolayers due to technical difficulties in measuring fluorescent dye signals in intact tissue. However, we now have the ability to monitor Ca^{2+} changes in the intact organ or perfused tissue, providing a unique opportunity to examine $[Ca^{2+}]_i$ regulation *in situ* while maintaining the normal geometry and coupling between cells. Studies in a variety of tissue preparations (12–21) which preserved the functional integrity of the intact tissue, and in perfused organs (4, 22–28) have revealed that propagation of the intracellular $[Ca^{2+}]_i$ wave is not limited to the initiating cell, but rather the Ca^{2+} signals can be communicated between neighbouring cells giving rise to intercellular $[Ca^{2+}]_i$ waves. The propagation of IP_3-dependent intercellular $[Ca^{2+}]_i$ waves provides a mechanism to regulate the function of a large number of cells in a coordinated fashion.

In our studies of the perfused rat liver, infusion of submaximal doses of IP_3-producing agonists initiated coordinated intercellular Ca^{2+} waves, which propagated across entire lobules and reached distances of up to 1 mm (4, 22–24). These intercellular Ca^{2+} waves initiated from a small number of periportal hepatocytes located near the portal vein, and then propagated in a radial fashion into the portal tract and outward to the pericentral region (please see Plate 9). When examined with sufficient spatial and temporal resolution, intracellular and intercellular $[Ca^{2+}]_i$ waves were easily observed propagating from cell to cell along the hepatic plates (please see Plate 10). These complex Ca^{2+} signals appear to be mediated by gap junctional communication with either IP_3 and/or Ca^{2+} as the propagating Ca^{2+} mobilizing messenger (4, 21–24, 29–30). Moreover, this preparation is well-suited for the investigation of autocrine, paracrine, and/or neuronal input in regulating agonist-induced $[Ca^{2+}]_i$ signals. For example, we have recently reported that endothelial-derived nitric oxide sensitized IP_3-dependent $[Ca^{2+}]_i$ signals in adjacent hepatocytes in the intact perfused liver (24). Although, this type of intercellular communication can sometimes be replicated in a co-culture system, the physiological significance can only be substantiated in the intact tissue.

The development of imaging techniques to monitor $[Ca^{2+}]_i$ homeostasis *in situ* should greatly enhance our understanding of how integration of Ca^{2+} signals translates into regulation of tissue function. In this chapter, we will summarize some of the recent methodological advancements from our laboratory designed to measure $[Ca^{2+}]_i$ responses in single cells within the intact perfused liver. These techniques can be adapted to measure $[Ca^{2+}]_i$ and other parameters (with appropriate fluorescent dyes) in other perfused organs and superfused tissues.

2 Perfusion apparatus and imaging systems

A key requirement for isolating and imaging an intact organ on a microscope stage is a mechanism to maintain oxygenation and supply of substrates and effectors. Apart from very small tissues or at least tissues with a small cross-sectional area where superfusion may be sufficient, some kind of perfusion apparatus is necessary. A suitable specimen chamber is also needed to stabilize the preparation on the microscope stage. In this section, we will describe the equipment currently used in our laboratory for liver perfusion. However, the design can be modified to meet the needs of a variety of tissues and organs.

2.1 Perfusion system

Intact tissues can be perfused by either gravitational flow or with a perfusion pump. The choice of perfusion system is dependent on the tissue preparation and experimental design. Gravity-fed perfusion maintains a constant pressure with the flow rate determined by vascular resistance of the tissue. This type of perfusion apparatus minimizes changes in shear stress and gives less tissue movement than a perfusion pump that may yield pressure fluctuations. A perfusion pump generates constant media flow and is easier to install and maintain. However, constant flow perfusion increases tissue pressure and may alter the function of certain cell types. For example, in our studies on the intercellular communication between endothelial cells and hepatocytes in the intact liver (24), we chose to use a gravity-flow perfusion system because endothelial cells may be activated by changes in vascular pressure (31).

The overall configuration and components of our constant flow perfusion system are schematically depicted in *Figure 1*. Most of the materials required to construct this apparatus should be readily available from suppliers such as Cole-Parmer and Fisher Scientific. The basic set-up consists of a peristaltic pump (used during the initial phases of perfusion and dye loading), an artificial lung for oxygen exchange, and a mixing chamber with magnetic stirrer for infusion of test agents. For gravitational flow, a line can be inserted into the perfusion system through a three-way stopcock placed just prior to the membrane lung (*Figure 1*); Tygon tubing and fittings connect the three-way stopcock to a buffer reservoir placed above the microscope stage (not shown). The desired flow rate can be obtained by adjusting tubing size and height of the reservoir.

The perfusion line between the membrane lung and the buffer reservoir consists of Tygon tubing (Formulation R-3603: 3.2 mm × 4.8 mm × 0.8 mm). The size

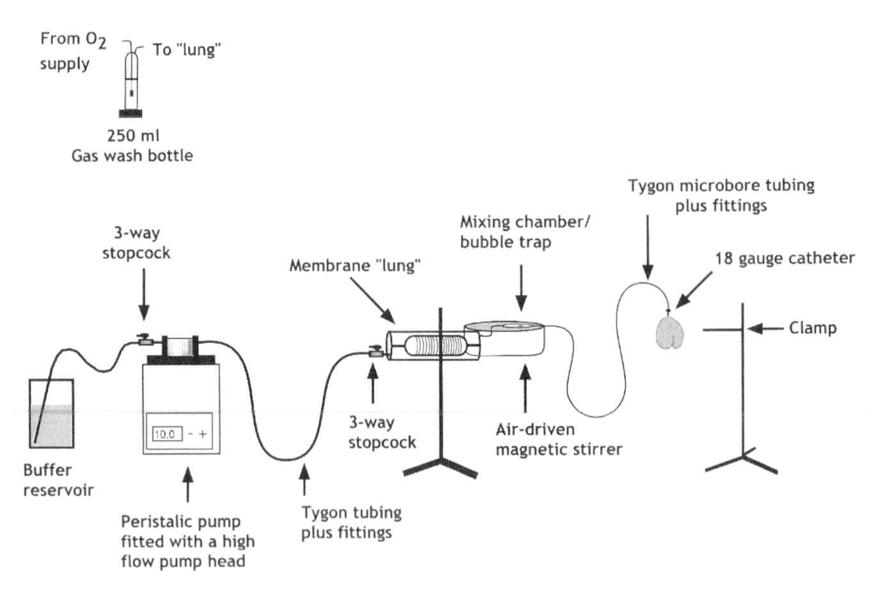

Figure 1 Schematic of a perfusion apparatus.

of the Tygon tubing in the pump head manifold is 2.4 mm × 4.0 mm × 0.8 mm and held in place with reducing connectors. This set-up generates a maximal flow rate of 25 ml/min, however the size of the tubing can be varied to obtain any desired flow rate. A three-way stopcock is also inserted into the perfusion line on the inlet side of the peristaltic pump; allowing the investigator to switch between the buffer reservoir (*Figure 1*) and a secondary buffer source that we use to supply the dye loading buffer (not shown). After dye loading, this second line can be switched to a buffer reservoir containing experimental compounds. The peristaltic pump (e.g. Gilson Minipuls 3) is fitted with a high flow pump head that has ten rollers, which helps to dampen pulsatile flow inherent with these pumps. Note the pulsation in the perfusion flow alters the image focus plane. Reducing fittings and Tygon microbore tubing (Formulation S-54-HL: 1.27 mm × 2.28 mm × 0.51 mm) connect the membrane lung to the mixing chamber and then to the catheter.

2.2 Membrane lung

Tissue oxygen deficiency interferes with normal function and therefore efficient oxygenation is critical for successfully imaging an intact organ. There have been many oxygenation chambers described in literature that are suitable for perfusing intact organs, a full description of which can be found in refs 32 and 33. We have modified the design of the oxygenation chamber so that it is compact enough to fit next to the microscope and imaging system without affecting their operation. In this section, the integral parts and configuration of our latest membrane 'lung' will be described.

Figure 2 schematically illustrates the size and dimensions of the components used to construct our membrane 'lung'. The oxygenation chamber consists of an

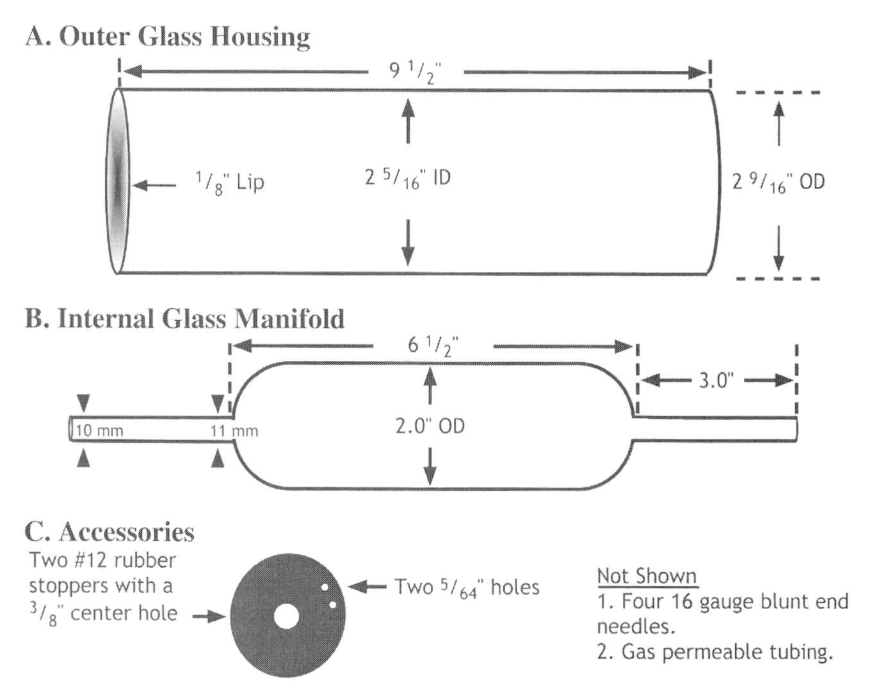

Figure 2 Components of an artificial lung.

outer glass housing (*Figure 2A*), an internal glass manifold (*Figure 2B*), two No 12 rubber stoppers, gas permeable tubing and fittings (*Figure 2C*). Local glassblowers were contracted to custom manufacture the glass components (The Glassblowers, Inc. PO Box 8089, Turnersville, NJ 08012; Tel: (609) 232-7898), while the rubber stoppers, gas permeable tubing and fittings were obtained from local laboratory supply distributors. The gas permeable tubing is Dow Cornings' Silastic brand tubing. The dimensions of the Silastic tubing are critical for adequate gas exchange; therefore, the oxygen content of the media effluent from the perfused organ should be measured. In addition, the mitochondrial flavoprotein fluorescence (excitation 470 nm, > 520 nm emission) can be used as an index of tissue oxygenation. For example in the intact liver, an adequately perfused and oxygenated tissue has a greenish-yellow fluorescence when excited at 470 nm, while anoxic tissue has a reddish-brown fluorescence.

Protocol 1

Assembling a membrane lung

Equipment

- Inner and outer glass components obtained from local glassblowers (e.g. The Glassblowers, Inc.)
- Two No 12 rubber stoppers
- Four 16 gauge blunt-end needles
- 1.47 mm × 1.95 mm × 0.24 mm Silastic brand tubing (e.g. Fisher Scientific)

Protocol 1 continued

Method

1. Bore an approx. three-eighths inch hole into the centre of two No 12 rubber stoppers. The holes should be sufficiently large so that the stoppers can fit snugly onto the ends of the internal glass manifold.
2. Drill two approx. 5/64th inch holes along the edge of each stopper.
3. Insert an inch long 16 gauge blunt-end stainless steel needle into each hole. The needles should also fit snugly.
4. Slide one of the stoppers approx. 1.5 inches onto the internal manifold.
5. Cut two pieces of Silastic brand tubing approx. 25 feet (7.5 metres) in length.
6. Connect to the two 16 gauge needles; securing the tubing to the needle with silk suture.
7. Anchor the tubing onto the internal manifold with tape.
8. Wrap the tubing around the manifold and anchor on the other side.
9. Insert internal manifold into the outer glass housing until the rubber stopper and housing form a seal.
10. Slide the second rubber stopper partly onto the internal manifold.
11. Secure the Silastic tubing onto the needles in this stopper.
12. Slide the second rubber stopper onto the internal manifold until it forms a seal with the outer housing. The stoppers should hold the internal manifold away from the inner surface of the glass housing while forming an airtight seal.
13. Insert a 1.5 inch 18 gauge needle through each rubber stopper. These will be the inlet and outlet ports for oxygen supply.
14. Mount the finished oxygenation chamber onto a support stand.
15. Connect the internal manifold to a circulating water-bath.

This set-up provides two separate perfusion lines, although only one is used during an imaging experiment. The second line is used as a back-up or when using perfusion buffers containing hydrophobic dyes or drugs that maybe difficult to remove from the tubing. The different inlets and outlets are colour-coded to avoid confusion.

2.3 Mixing chamber/bubble trap

The mixing chamber/bubble trap is constructed from a rubber flashball device and plastic needle adapter taken from an intravenous drip chamber (Baxter basic solution set). The flow from the needle adapter passes into a 3.5 cm piece of Tygon tubing (Formulation: R-3603, 4.8 mm × 7.9 mm × 1.6 mm) containing a 7 mm × 2 mm magnetic stir bar. This stir bar is driven by an air turbine magnetic stirrer (Fisher Scientific) to ensure mixing of the infusate. We use an air-driven stirrer to limit the electrical fields generated close to the imaging equipment. Reducing fittings and Tygon microbore tubing (1.27 mm × 2.28 mm

× 0.51 mm) connect the flashball device to the membrane 'lung' and to an 18 gauge Teflon intravenous catheter (Becton Dickinson). Agonists or drugs are infused directly into the mixing chamber through a 1 inch 23 gauge needle inserted into the flashball device. A 5 cc syringe is attached to the 23 gauge needle with reducing fitting and microbore tubing. The infusion rate is controlled by a syringe pump and held between 0.4–1.0% of perfusion flow rate.

2.4 Imaging chamber

The final step is the design of an imaging chamber, which provides support for the perfused organ and prevents tissue movement during data acquisition. Movement is particularly problematic during agonist stimulation, since many of the hormones that regulate liver physiology (e.g. α-adrenergerics, eicosanoids, glucagon, purines, and vasopressin) also alter vascular tone and/or cell volume (34–36). The end-result can be a dramatic alteration in perfusion pressure that causes tissue movement affecting the x, y position and the plane of focus. While motion artefacts give rise to obviously spurious fluorescence changes with ratiometric dyes, they can often be misinterpreted as Ca^{2+} changes with single wavelength Ca^{2+}-sensitive indicators.

A simple imaging chamber can be assembled from the tops or bottoms of plastic tissue culture dishes whose final dimensions are determined by the design of the microscope stage. However, larger chambers allow easier handling and positioning of the perfused organ. An example of an imaging chamber currently used in the authors' laboratory is illustrated in *Figure 3*. This chamber is constructed from the top of a Nunclon 8-well multidish.

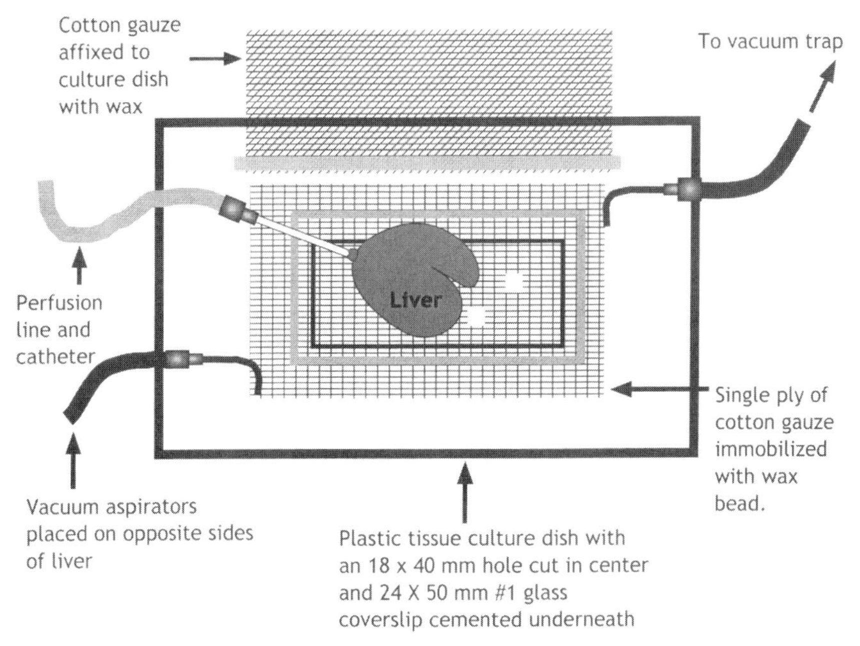

Figure 3 Schematic of an imaging chamber used to support and immobilize perfused tissues.

Protocol 2

Assembling an imaging chamber

Equipment

- Plastic Petri dish
- Plastic cutting tool, e.g. soldering iron
- Silicone adhesive sealant, e.g. Permatex black silicone
- 24 mm × 50 mm No 1 glass coverslip
- Cotton gauze
- Paraffin wax, e.g. tissue embedding media (Fischer Scientific)

Method

1. Cut a hole (ca. 18 mm × 40 mm) into the plastic dish.
2. Sand the edges surrounding the hole to remove burrs.
3. Trim the sides of the culture dish. This allows easier access for cannula and perfused tissue.
4. Use a silicone adhesive sealant to glue the glass coverslip onto the bottom of the Petri dish. Note that this should form a watertight seal.
5. Allow the sealant to cure.
6. Secure a single ply of cotton gauze on top of the coverslip with a bead of melted wax. This layer of gauze prevents the tissue from sliding around on the coverslip.
7. Scrape and level the sides of the wax bead to obtain a flat surface.
8. Cut several small holes in the cotton gauze. These holes provide large, relatively unobstructed 'windows' to image the perfused organ.
9. Affix a 12-ply, 4″ × 4″ gauze sponge to one edge of the culture dish with wax. The gauze sponge folds down over the top of the tissue holding it flat against the coverslip.

Effluent is drawn into the gauze by capillary action and rapidly removed by placing vacuum aspirators on either side of the tissue. In our hands, media accumulation between the liver and the glass coverslip exacerbates movement. Bending 13 gauge needles at approximately 45° angle makes the aspirator tips.

2.5 Imaging devices

The advantages and disadvantages of the currently available Ca^{2+} imaging devices has been discussed in detail (37–39). In our intact perfused liver Ca^{2+} studies, we use two different types of imaging systems depending on the type of information required. A laser scanning confocal system has been used to record images with high spatial resolution and to determine the intercellular $[Ca^{2+}]_i$ wave pathway (please see Plates 9 and 10). This system also has a sufficient temporal resolution to monitor the frequency of $[Ca^{2+}]_i$ oscillations. However, since our confocal microscope system is not equipped with the UV excitation that is required for ratiometric Ca^{2+} indicators, we have also utilized a charge coupled device (CCD) camera in conjunction with conventional epifluorescence microscopy to obtain quantitative $[Ca^{2+}]_i$ measurements. This system is suitable for acquiring additional information about the relationship between $[Ca^{2+}]_i$ oscillations and mitochondrial pyridine nucleotides (see Section 6.2). Note that images obtained from intact tissues with conventional epifluorescence measurements are heavily contaminated with out-of-focus fluorescence signals and require post-acquisition image deconvolution (see Section 6.2).

An inverted IMT-2 Olympus microscope is attached to Bio-Rad MRC 600 laser scanning confocal system equipped with a standard argon ion laser and manufacturer-supplied data acquisition software (CoMOS) and filter sets. Fluo-3 fluorescence images are obtained using a standard fluorescein filter set consisting of 488 DF 10 excitation filter, 510 LP dichroic reflector, and OG 515 LP emission filter. Laser intensity is attenuated to 0.3–0.5% transmission by placing additional neutral density filters (2.3–2.5 o.d.) into the light path. Clearly, proper alignment of the laser light path is crucial when using these low-level excitation intensities. A Zeiss Plan-NEOFLUAR 0.5 NA $\times$ 16 objective or a Nikon UV-F 0.8 NA $\times$ 20 objective is used to collect low magnification images, while an Olympus SPlan Apo 1.3 NA $\times$ 60 objective is used to obtain images of high resolution. Because of concerns regarding signal loss due to photobleaching of the dye, the standard data acquisition parameters are designed to limit the amount of laser excitation. In addition, exposing the liver to long periods ($>$ 20 min) of continuous laser scanning can damage the tissue and disrupt intercellular $[Ca^{2+}]_i$ waves, although this can be reversed by a short recovery period provided that the laser intensity and/or total exposure is not excessive. In our system, the level of tissue exposure to laser light can be reduced and the rate of acquisition increased by using a subframe image size (e.g. 255 $\times$ 170 pixels) with a low electronic zoom (1.0–1.3). Signal-to-noise can be enhanced by averaging multiple frames during data acquisition. In the intact liver, intercellular $[Ca^{2+}]_i$ waves take several minutes to cross the entire lobule, therefore a slow data acquisition rate (one image/3–4 sec) is sufficient to monitor these Ca^{2+} waves. The photomultiplier gain should be set to maximize the peak signal without saturation. The size of the confocal aperture or 'pin-hole' is critical for acquiring spatially resolved $[Ca^{2+}]_i$ signals. Images obtained with a large optical section have a higher total signal, however the fluorescence is likely to arise from more than one cell. In the liver, we routinely use a confocal aperture, which translates, into an optical section of 20–40 μm, but smaller apertures may be appropriate for tissues where the cells are smaller.

The CCD-based system consists of a cooled slow scan CCD camera attached to an inverted Nikon Diaphot TMD microscope. We again use relatively small image sizes (128 $\times$ 96 pixels or 192 $\times$ 144 pixels) and pixel binning to improve signal-to-noise and increase read-out rate. Data acquisition and the selection of excitation filters and exposure times (200–250 msec) is computer controlled through a Ludel/Uniblitz filter wheel/shutter assembly using custom software. Fura-2 fluorescence images are obtained using 340 DF 13 and 380 DF 13 excitation filters, a 400 DCLP dichroic reflector, and a 420 LP emission filter. During data acquisition, the xenon light source is attenuated with a 2.1 o.d. neutral density filter. Fura-2 fluorescence is collected using fluorescence filter sets purchased from either Chroma Technology Corporation or Omega Optical in conjunction with either a Nikon UV-F 0.8 NA × 20 objective or a Zeiss Plan-NEOFLUAR 1.25 NA × 63 objective.

3 Perfusion

A viable tissue preparation is ultimately the most important criterion governing the success or otherwise of imaging in the intact liver. In addition to compromised function, traumatized or poorly perfused livers compartmentalize significant quantities of Ca^{2+} indicator dyes into subcellular organelles which interferes with calibration. Even localized tissue damage can sufficiently slow or abrogate circulation in distal parts of the organ. Thus, it is advisable to discard any preparation that shows signs of damage or inadequate perfusion. In the intact liver, white blotches on the tissue surface are a sign of damage during isolation. Adequate tissue oxygenation can also be assessed by monitoring the mitochondrial flavoprotein fluorescence as discussed in Section 2.2.

Whilst this section will describe the techniques used in our laboratory to isolate and perfuse an intact rat liver, the handling of various other perfused tissues has been described elsewhere (33). The reader can also refer to previous literature describing single cell $[Ca^{2+}]_i$ imaging studies in the perfused rat lung (27), pyramidal neurons in rat brain (40), and perfused rat heart (28). In addition, several elegant studies have monitored single cell $[Ca^{2+}]_i$ responses in multicellular preparations which maintain the functional unit of the intact tissue including: blowfly salivary glands (12, 13); hippocampal slices (14, 15); rat tail artery (16); tracheal epithelial cells (17, 18); retina (19); pancreatic acini clusters (20, 21).

3.1 Perfusion media

The choice of appropriate perfusion media will depend on the tissue used and the duration of experiment. We use a Hepes-buffered balanced salt solution for our intact liver imaging studies (22, 24). This simple medium maintains a viable tissue preparation for up to three hours, although there is a slight decrease in agonist sensitivity over time. The final intact perfused rat liver (IPRL) medium composition is shown in *Protocol 3*. We routinely prepare 4–6 litres of IPRL medium fresh each experimental day. To limit the amount of dye required, the liver is loaded with Ca^{2+} indicator dyes by recirculating the same medium with the BSA increased to 2% (w/v) (see Section 4.1). In experiments using fura-2 loaded livers, autofluorescence images are obtained post-experimentally using an *in situ* calibration medium, which is Ca^{2+}-free IPRL medium plus 0.1 mM $MnCl_2$. 0.5 litres of the IPRL medium is set aside prior to the addition of $CaCl_2$ to make up the *in situ* calibration media (see Section 5.1). All media is filtered through a 0.8 μm membrane filter prior to use to remove debris.

3.2 Tissue isolation

Our $[Ca^{2+}]_i$ studies in the intact perfused rat liver have centred upon the median lobe. The whole rat liver can be used for $[Ca^{2+}]_i$ imaging studies, however we prefer a single lobe for several reasons:

(a) A single liver lobe requires less Ca^{2+} indicator dye for adequate loading.

(b) Slower flow rate decreases the volume of perfusate needed.

(c) Single lobes are easier to manipulate and immobilize on a microscope stage.

(d) IP_3-dependent intracellular and intercellular $[Ca^{2+}]_i$ waves are qualitatively similar in the whole liver and single lobes.

An excellent description of the perfused rat liver technique can also be found in ref. 33.

Protocol 3

Liver perfusion technique

Equipment and reagents

- 4–6 litres of tissue perfusion media: e.g. 121 mM NaCl, 25 mM Hepes, 5 mM $NaHCO_3$, 4.7 mM KCl, 1.2 mM KH_2PO_4, 1.2 mM $MgSO_4$, 1.3 mM $CaCl_2$, 0.25 mM bromosulfopthalein (BSP),[a] 5.5 mM glucose, 0.5 mM glutamine, 3.0 mM lactate, 0.3 mM pyruvate, 0.1% (w/v) dialysed bovine serum albumin (BSA) pH 7.4, equilibrated with 100% oxygen at 30 °C
- Tissue perfusion apparatus (see Section 2)
- Small animal surgical equipment, e.g. operating scissors, forceps, and micro-dissecting scissors
- 2-0 surgical silk
- Anaesthetic, e.g. Nembutal

Method

1. Prime the perfusion apparatus with buffer and start recirculating the tissue perfusion media.
2. Expel all air bubble from the perfusion lines.
3. Anaesthetize a 200–300 g male Sprague–Dawley rat with 60 mg Nembutal/kg body weight.
4. Place the anaesthetized animal on a mobile operation stage.
5. Open the abdominal cavity widely using a curved operating scissors.
6. Move the stomach and intestines to the left side exposing the portal vein and inferior vena cava.
7. Arch the animal's back; forcing the median and left lobes against the diaphragm and thus separating the lobes from the portal vein.
8. Cut all connective tissue between liver lobes.

Protocol 3 continued

9 Place three loose ligatures (surgical silk) around the portal vein:
   (a) Near its entry into the liver.
   (b) Between the splenic vein and mesenteric arteries.
   (c) Distal to the second ligature.
10 Adjust the perfusion flow rate to approx. 10–15 ml/min.
11 Apply gentle traction to the third ligature.
12 Using a curved micro-dissecting scissors, make a small incision, at an angle of 45°, in the portal vein.
13 Insert and thread the IV catheter into the portal vein. The entire liver should immediately blanch and become a homogeneous brown colour. The catheter may have to be retracted slightly to allow the blood to clear from the caudate and right lobes.
14 Secure the catheter by tying off the top two ligatures.
15 Sever the inferior vena cava while increasing the perfusion flow rate to 25 ml/min.
16 Remove rib cage exposing the chest cavity.
17 Sever the aorta.
18 Move the median and left lobes up into the chest cavity and place them on a piece of cotton gauze.
19 Place loose ligatures around the left, right, and caudate lobes.
20 Quickly tie off and excise the lobes. Note: When the left lobe is tied off it also occludes the right portion of the median lobe, but this does not affect $[Ca^{2+}]_i$ measurements in the rest of the median lobe.
21 Remove the median lobe and hang on the three-prong clamp.
22 Rinse the lobe with normal saline to remove any blood clots on the surface.
23 Check for tissue damage (i.e. visually or by mitochondrial flavoprotein fluorescence).
24 Allow the liver to recover for 10–15 min prior to loading with Ca^{2+}-sensitive indicators.
25 With the exception of the dye loading period, the liver is perfused in a single pass mode.

[a] BSP is added to the media to inhibit dye loss, see Section 4.2.

4 Loading Ca^{2+} indicator dyes

4.1 Loading protocol

Ca^{2+}-sensitive indicator dyes are loaded by recirculating 60 ml of IPRL media containing acetoxymethyl ester (2–5 μM) plus 0.01% (w/v) Pluronic acid F-127 and 2% BSA (w/v) for 20–30 min. When simultaneously measuring $[Ca^{2+}]_i$ and mitochondrial pyridine nucleotide responses in the intact liver, it may be necessary to decrease the concentration of the fura-2 AM (48, 49). The dye loading process can be visualized when using visible wavelength Ca^{2+} indicators (e.g. fluo-3) by monitoring the decrease in the colour of the loading buffer and its appearance in the tissue. Dye loading is terminated when the buffer becomes clear.

Protocol 4

Ca^{2+} indicator loading protocol

Reagents

- Tissue perfusion media (see *Protocol 3*)
- 10% (w/v) Pluronic acid F-127 solution in dry DMSO
- Acetoxymethyl ester of Ca^{2+} indicator at 1 mM in dry DMSO
- 18% (w/v) dialysed bovine serum albumin (BSA) solution

Method

1. Pipette 54 ml of perfusion media into a 100 ml graduated cylinder.
2. Add 6 ml of BSA solution.
3. Mix 300 μl of the acetoxymethyl ester dye stock with 60 μl of Pluronic acid F-127. The final indicator-AM concentration is 5 μM, while Pluronic F-127 is 0.01% final.
4. Add dye/Pluronic acid solution to the loading buffer.
5. Mix thoroughly.
6. Recirculate dye-containing media for 20–30 min.

4.2 Dye retention

The intact liver is very efficient at extracting organic fluorescent markers from the circulation and secreting them into the bile. While this has proved advantageous in the study of bile canalicular motility and bile flow (41), hepatobiliary secretion is a significant drawback to efficiently loading Ca^{2+}-sensitive indicators. *Figure 4* shows the tissue distribution of fluo-3 fluorescence in the absence of any manipulations to block bile canaliculi secretion: most of the hydrolysed free acid form of the dye was rapidly extruded into the canalicular lumen as indicated by the polygonal pattern of fluorescence and the absence of any nuclear signal. The loss of dye into the bile canaliculi can be slowed by decreasing the perfusate temperature to 30 °C and adding bromosulfopthalein (BSP) to the media. BSP is a competitive organic anion transport inhibitor, which does not affect liver viability and is used clinically to assess liver function (42). Moreover, BSP does not affect IP_3-dependent $[Ca^{2+}]_i$ responses when added to isolated rat hepatocytes (Gaspers and Thomas, unpublished results). BSP was the only transport inhibitor tested that effectively blocked Ca^{2+} indicator dye loss in the intact liver. Note that BSP-containing media are a deep purple colour at neutral to alkaline pH, but BSP is not fluorescent and does not affect fura-2 or fluo-3 signals.

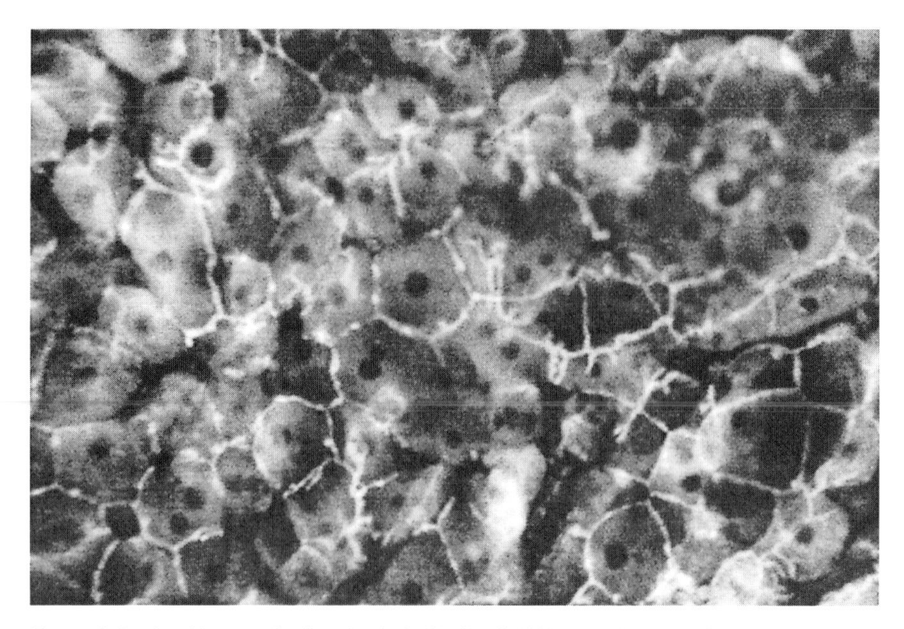

Figure 4 Confocal image of a liver loaded with fluo-3 AM in the absence of the organic anion transport inhibitor, BSP. Note the intense labelling in the lumen of the bile canaliculi.

4.3 Photobleaching

The intact perfused liver has an intense and extremely photolabile yellowish-green autofluorescence when excited in the UV range, used for fura-2 measurements, which has been suggested to originate from Kupffer cells (43). In contrast, a photosensitive autofluorescence component is not observed during visible wavelength excitation. The UV-sensitive autofluorescence tends to obscure the underlying liver structure and contributes unevenly to the 340 nm and 380 nm excitation making it difficult to calibrate $[Ca^{2+}]_i$ measurements. Since this component bleaches rapidly, we deliberately pre-bleach it by collecting 50–100 fluorescence images using a low optical density neutral density filter prior to $[Ca^{2+}]_i$ measurements. As the photolabile component disappears, the liver structure becomes more defined. We have confirmed that this controlled photobleaching period does not affect tissue viability or subsequent $[Ca^{2+}]_i$ measurements. Apart from this initial pre-bleaching protocol, both the xenon lamp and argon laser excitation sources are highly attenuated to minimize photocytotoxcity. Strong excitation intensities can degrade Ca^{2+} indicator dyes and disrupt intercellular $[Ca^{2+}]_i$ waves.

5 Calibration

5.1 Calibration protocol

Calibration of fluorescence intensities into absolute $[Ca^{2+}]_i$ values relies on the same principles and is fraught with the same problems as single isolated cells

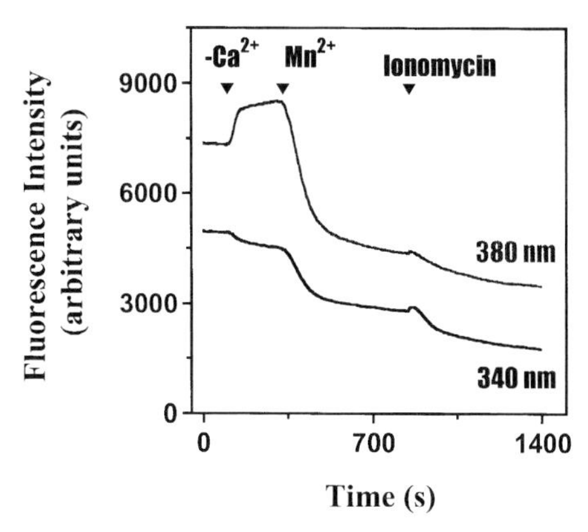

Figure 5 Calibration of fura-2 fluorescence intensities in the intact liver. Autofluorescence values plus compartmentalized fura-2 were determined at the end of the experimental run by sequential perfusion with a calcium-free media ($-Ca^{2+}$) followed by the same media plus 100 μM $MnCl_2$ (Mn^{2+}). Ionomycin (4 μM) infusion quenches the remaining fura-2 dye molecules trapped in the internal compartments. Note the fluorescence remaining after Mn^{2+} quench of cytosolic fura-2 is composed of cellular autofluorescence (e.g. NAD(P)H) plus compartmentalized fura-2; both values must be subtracted from Ca^{2+}-dependent signals. Therefore, the fluorescence just prior to ionomycin application is subtracted from all previously acquired images before calculating image ratios.

(37–39, and Chapter 2). In the intact liver, calibration requires an accurate determination of the tissue autofluorescence and the fluorescence contribution from Ca^{2+} indicator dye trapped in internal organelles. In fura-2 loaded livers, both parameters are determined following quench of the cytosolic dye with $MnCl_2$ at the end of the experiment. Since $MnCl_2$ does not totally quench the fluorescence of the currently available visible wavelength dyes, we have not attempted to calibrate these fluorescence data into absolute $[Ca^{2+}]_i$ changes. The calibration protocol is shown in *Figure 5*. In this example, we perfused the liver with Ca^{2+}-free media followed by the same media containing $MnCl_2$ as illustrated. These steps are usually combined to limit the time that the liver is exposed to Ca^{2+}-free media, since Ca^{2+} depletion causes the cells to dissociate from the extracellular matrix. It is also not necessary to use ionomycin to quench the remaining Ca^{2+} indicator dye in the internal compartments. The procedure for this type of calibration is detailed in *Protocol 5*. The R_{min}, R_{max}, and S_{f2}/S_{b2} values for fura-2 were determined *in vitro* at 30 °C as described (37). A K_d value of 184 nM is assumed at 30 °C.

Protocol 5

Fura-2 in situ calibration protocol

Reagents

- Ca^{2+}-free tissue perfusion media containing 100 μM $MnCl_2$ (see *Protocol 3*)

Protocol 5 continued

Method

1 Wash out the agonist or experimental drug for 20–30 min.
2 Switch to a Ca^{2+}-free perfusion media containing $MnCl_2$.
3 Monitor the fluorescence intensity changes until the 340 nm and 380 nm signals stabilize. In *Figure* 5, this is just prior to ionomycin addition.
4 Determine if any tissue movement has occurred during the calibration protocol. The position of user-defined ROIs before and after the calibration protocol is used to assess tissue movement.
5 If necessary, realign the tissue to its original position.
6 Acquire the autofluorescence images, which can then be subtracted from previously acquired images at the same wavelengths before calculating image ratios.

6 Image acquisition and analysis

6.1 Image quality

High quality images, even at low magnification, can be acquired by adjusting the tissue on the imaging chamber until a large area of cells (1 mm^2) are in a similar focal plane. In essence, we spread the liver tightly against one of the imaging 'windows' cut into the cotton gauze that is secured to the coverslip (see *Figure 3*): a forceps handle is inserted underneath the liver lobe and then gently pulled forward and upward. A second piece of cotton gauze is placed over the flattened liver and held in place with the effluent aspirators and adhesive tape. It maybe necessary to repeat this procedure several times before the liver is properly immobilized. Despite these precautions, unevenness in the liver surface contributes to differences in the absolute fluorescence intensities across the lobule. However, this should not affect the magnitude of normalized $[Ca^{2+}]_i$ spikes measured with fluo-3 or the calibrated $[Ca^{2+}]_i$ values obtained from deconvolved fura-2 images (see Section 6.2).

6.2 Deconvolving non-confocal fura-2 images

The maximum useful penetration of UV or visible excitation light into the intact tissue is approximately 100 μm when using a low magnification objective. Therefore, in conventional epifluorescence measurements the image is contaminated with a substantial amount of signal from out-of-focus cells (see *Figure* 6). This overlap of single cell signals tends to broaden the $[Ca^{2+}]_i$ spike, slow the apparent kinetics and decrease the apparent amplitude of $[Ca^{2+}]_i$ change (see *Figure 7*). However, this problem can be overcome by confocal microscopy or by post-acquisition signal deconvolution (deblurring). With all the advantages conferred by ratiometric recording, it is unfortunate that Indo-1 is the only dual emission ratiometric Ca^{2+} indicator dye compatible with confocal microscopy. Indo-1

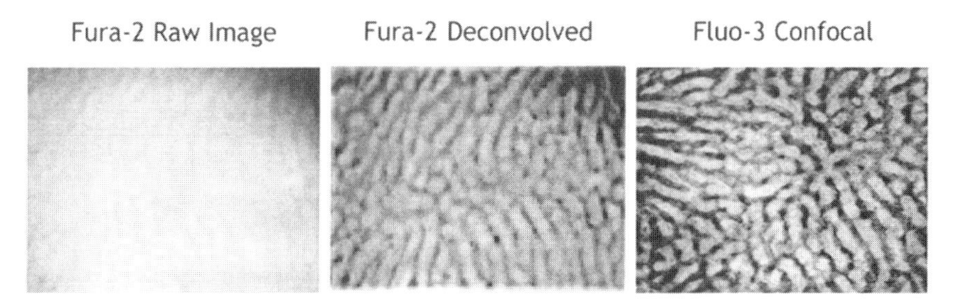

Figure 6 Comparison of deconvolved epifluorescence images with confocal microscopy. Epifluorescence images were acquired from a fura-2 loaded intact liver (*left*) and deconvolved as described in the main text (*middle*). (*Right*) An equivalent confocal image of a fluo-3 loaded liver.

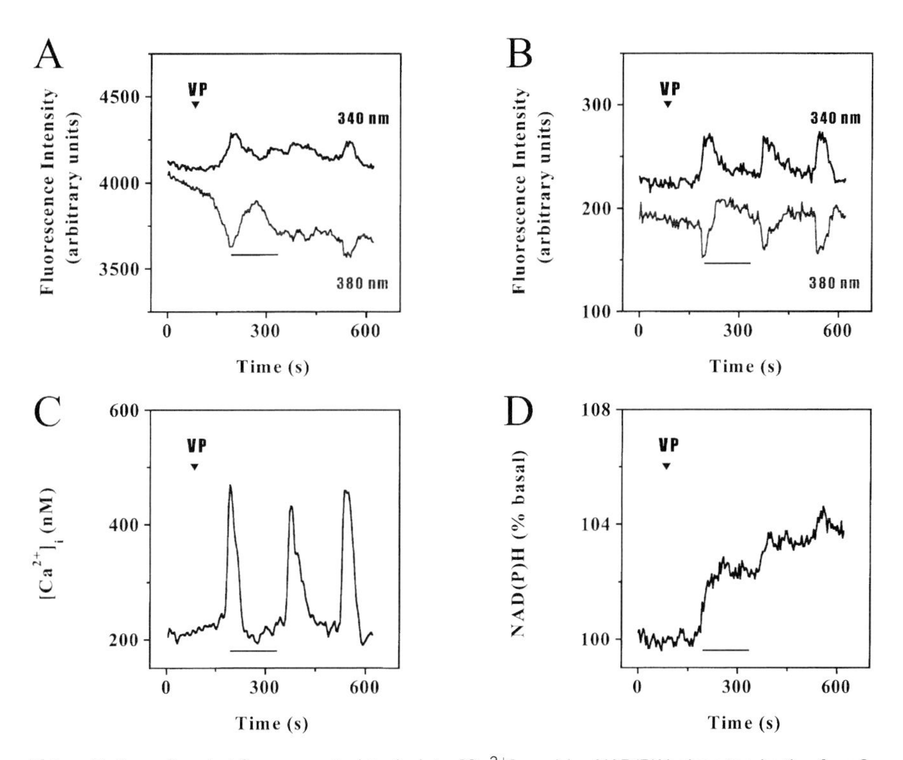

Figure 7 Transforming fluorescent signals into $[Ca^{2+}]_i$ and/or NAD(P)H changes in the fura-2 loaded liver. (A) The raw 340 nm and 380 nm fluorescence intensity changes recorded in a single hepatocyte from the intact perfused rat liver stimulated with vasopressin (VP). (B) The same 340 nm and 380 nm fluorescence intensity changes after deconvolution. (C) Calibrated $[Ca^{2+}]_i$ responses of panel B. (D) The change in NAD(P)H calculated from the 340 nm and 380 nm fluorescence intensities using the equation $y = \Delta A^*F_{340\ nm} + \Delta B^*F_{380\ nm}$. The constants ΔA and ΔB are empirically determined from the fractional fluorescence intensity changes at 340 nm and 380 nm during an increase in $[Ca^{2+}]_i$. Alternatively, NAD(P)H changes can be recorded by monitored 360 nm excitation in addition to the 340 nm and 380 nm (48, 49). The arrow head in each panel shows when VP reaches the liver and the time-line bar in each panel shows the onset of the NAD(P)H response.

requires either an UV laser source for excitation or a high powered visible light laser for two-photon excitation. In the long run it is likely that the two-photon approach will be best for measuring single cell $[Ca^{2+}]_i$ in thick tissues.

An alternative, less expensive approach to confocal microscopy is to use deconvolution of conventional microscopy images. A number of deconvolution algorithms have been developed using images from multiple focal planes or images stacks (44, 45). However, there is a substantial penalty in the time resolution when real time image acquisition must be performed in more than one focal plane. To overcome this problem we have used a single plane deblurring/deconvolution method based on the algorithms and formulae described by Monck *et al.* (46) and Castleman (47). Our software version of this approach can be applied to a single plane time series of images and yield sufficient removal of out-of-focus information to permit calibration of single cell $[Ca^{2+}]_i$ signals from conventional epifluorescence images of fura-2 loaded livers. A copy of the mathematical support routines written in C programming code can be obtained by writing to the authors. The deconvolution software uses the in-focus image and a theoretical point spread functions (PSF) to estimate the out-of-focus information originating from the focal planes above and below the in-focus image. A fast Fourier transform function converts both the blurred digital image and the theoretical PSF into the frequency domain. The estimated out-of-focus fluorescence is then subtracted from the in-focus blurred image and the corrected image is converted back to the spatial domain. The software has several user-defined parameters to estimate the theoretical point spread functions, which are determined empirically. This should be used judiciously however, since more stringent settings sharpen the $[Ca^{2+}]_i$ responses at the expense of image quality and overall fluorescence intensity.

It is common to have the expected zero fluorescence intensity values in the sinusoidal spaces between hepatic plates after deconvolution (*Figure* 6). However, the fura-2 fluorescence intensity signals from the in-focus cell and adjacent cells are not completely separated by the deconvolution process. This is most clearly observed during high $[Ca^{2+}]_i$ oscillation frequencies, where baseline $[Ca^{2+}]_i$ values appear to be significantly elevated during the inter-spike period. A similar phenomenon is observed using confocal microscopy if a relatively thick optical slice is used to collect fluorescence images. Under these conditions, the $[Ca^{2+}]_i$ responses are derived from multiple cell layers. Deconvolution also increases the noise at the periphery of the image (*Figure* 6). Therefore, regions of interest (ROIs) should not be defined in close proximity to these areas.

The power of the deconvolution software is illustrated in *Figure* 7. In the intact liver, vasopressin-stimulated $[Ca^{2+}]_i$ oscillations are associated with a concomitant NAD(P)H increase elicited by mitochondrial Ca^{2+} uptake, which has similar kinetics to the NAD(P)H responses observed in isolated hepatocytes (48, 49). IP_3-dependent increases in NAD(P)H decay more slowly than the $[Ca^{2+}]_i$ transient resulting in a sustained mitochondrial response at low $[Ca^{2+}]_i$ oscillation frequencies (*Figure 7D*). The time-line bar in Figures 7A–C show the onset of the mitochondrial pyridine nucleotide response to vasopressin infusion in the intact liver. *Figures 7A* and *7B* show the 340 nm and 380 nm fluorescence intensity changes from the same hepatocyte before (A) and after deconvolution (B). In the raw fura-2 image data set, the fluorescence intensities change slowly (A), which would tend to broaden the $[Ca^{2+}]_i$ spike. In fact, the increase in mitochondrial pyridine nucleotide reduction almost completely obliterates the second $[Ca^{2+}]_i$ spike which is only revealed after deconvolution. *Figure 7C* shows the calibrated $[Ca^{2+}]_i$ change after deconvolution and subtraction of the autofluorescence values.

7 Future directions

The explosion of green fluorescent protein (GFP)-based indicators which are capable of monitoring a variety of cellular events from IP_3 formation (50) to cytosolic $[Ca^{2+}]_i$ changes (51) led us to explore the possibility of transiently transfecting perfused organs with plasmids encoding genes of interest. The whole rat liver was isolated and perfused with William's E media plus 5% (v/v) fetal calf serum, insulin (10 nM), and EGF (10 μg/ml). A plasmid encoding GFP was coated onto 1 μm gold beads and the coated beads 'shot' into the tissue using a helium gas pulse from a Bio-Rad Helios Gene Gun system as described by the manufacturer. Laser scanning confocal microscopy monitored GFP expression over time in terms of fluorescence, which was evident two hours post-transfection with no detectable cellular damage (*Figure 8*). The gold beads however, were easily identified in the tissue because the beads blocked laser excitation and thus, ablated the cellular autofluorescence or GFP fluorescence around the bead. In the right panel of *Figure 8*, the dark dots within the cells expressing GFP are the gold beads. This protocol can hopefully be optimized to transiently express different GFP–fusion proteins. Note that this transient transfection protocol was carried out in an *ex vivo* perfused tissue; performing local surgery then allowing the animal to recovery overnight might increase gene expression.

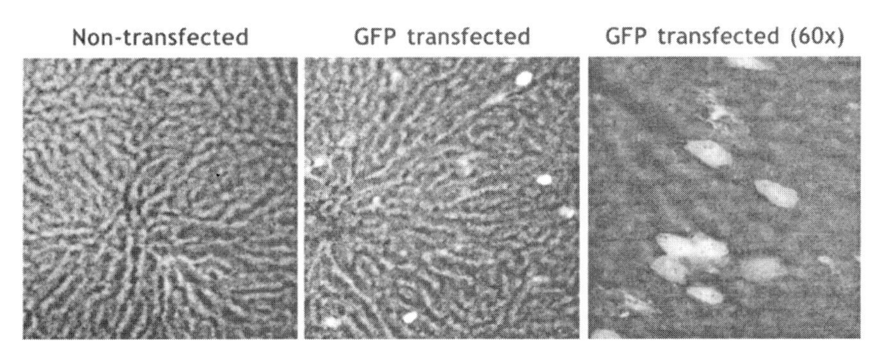

Figure 8 Transfection of an *ex vivo* perfused intact rat liver. A whole liver was isolated and perfused with William's E media plus 5% fetal calf serum, insulin (10 nM), and EGF (10 μg/ml). The liver was transfected with GFP using a Gene Gun. Images were acquired by confocal microscopy prior to transfection (*left*), two hours post-transfection (*middle*), and four hours post-transfection (*right*). The *left* and *centre* images were acquired using the same instrument settings whereas the *right* panel was acquired at greater magnification.

Acknowledgements

We thank Dr Paul Burnett for his technical assistance with GFP transfection of the intact liver and Dr Soraya S. Smaili and Basil M. Hantash for helpful discussions. This work was supported by National Institutes of Health Grant RO1 DK38422 (to A. P. T.) and a Wellcome Prize Traveling Research Fellowship (to S. P.).

References

1. Berridge, M. J., Bootman, M. D., and Lipp, P. (1998). *Nature*, **95**, 645.
2. Means, A. R., Vanberkum, M. F. A., Bagchi, I., Lu, K. P., and Rasmussen, C. D. (1991). *Pharmacol. Ther.*, **50**, 255.
3. Berridge, M. J. (1993). *Nature*, **361**, 315.
4. Thomas, A. P., Bird, G. St. J., Hajnóczky, G., Robb-Gaspers, L. D., and Putney, J. W. Jr. (1996). *FASEB J.*, **10**, 1505.
5. Patel, S., Joseph, S. K., and Thomas, A. P. (1999). *Cell Calcium*, **25**, 247.
6. Putney, J. W. Jr. and Bird, G. S. (1993). *Cell*, **75**, 199.
7. Woods, N. M., Cuthbertson, K. S., and Cobbold, P. H. (1986). *Nature*, **319**, 600.
8. Woods, N. M., Cuthbertson, K. S., and Cobbold, P. H. (1987). *Cell Calcium*, **8**, 79.
9. Rooney, T. A., Sass, E. J., and Thomas, A. P. (1989). *J. Biol. Chem.*, **264**, 17131.
10. Rooney, T. A., Sass, E. J., and Thomas, A. P. (1990). *J. Biol. Chem.*, **265**, 10792.
11. Hajnóczky, G. and Thomas, A. P. (1997). *EMBO J.*, **16**, 3533.
12. Zimmermann, B. and Walz, B. (1997). *J. Physiol.*, **500**, 17.
13. Zimmerman, B. and Walz, B. (1999). *EMBO J.*, **18**, 3222.
14. Dani, J. W., Chernjavsky, A., and Smith, S. J. (1992). *Neuron*, **8**, 429.
15. Harris-White, M. E., Zanotti, S. A., Frautschy, S. A., and Charles, A. C. (1998). *J. Neurophysiol.*, **79**, 1045.
16. Iino, M., Kasai, H., and Yamazawa, T. (1994). *EMBO J.*, **13**, 5026.
17. Sanderson, M. J., Charles, A. C., Boitano, S., and Dirksen, E. R. (1994). *Mol. Cell. Endocrinol.*, **98**, 173.
18. Salathe, M. and Bookman, R. J. (1995). *J. Cell Sci.*, **108**, 431.
19. Newman, E. A. and Zahs, K. R. (1997). *Science*, **275**, 844.
20. Stauffer, P. L., Zhao, H., Luby-Phelps, K., Moss, R. L., Star, R. A., and Muallem, S. (1993). *J. Biol. Chem.*, **268**, 19769.
21. Yule, D. I., Stuenkel, E., and Williams, J. A. (1996). *Am. J. Physiol.*, **271**, C1285.
22. Robb-Gaspers, L. D. and Thomas, A. P. (1995). *J. Biol. Chem.*, **270**, 8102.
23. Thomas, A. P., Robb-Gaspers, L. D., Rooney, T. A., Hajnóczky, G., Renard-Rooney, D. C., and Lin, C. (1995). *Biochem. Soc. Trans.*, **23**, 642.
24. Patel, S., Robb-Gaspers, L. D., Stellato, K. A., Shon, M., and Thomas, A. P. (1999). *Nature Cell Biol.*, **1**, 467.
25. Nathanson, M. H., Burgstahler, A. D., Mennone, A., Fallon, M. B., Gonzalez, C. B., and Sáez, J. C. (1995). *Am. J. Physiol.*, **269**, G1667.
26. Motoyama, K., Karl, I. E., Flye, M. W., Osborne, D. F., and Hotchkiss, R. S. (1999). *Am. J. Physiol.*, **276**, R575.
27. Ying, X., Minamiya, Y., Fu, C., and Bhattacharya, J. (1996). *Circ. Res.*, **79**, 898.
28. Hama, T., Takahashi, A., Ichihara, A., and Takamatsu, T. (1998). *Cell. Signal.*, **10**, 331.
29. Sáez, J. C., Connor, J. A., Spray, D. C., and Bennett, M. V. (1989). *Proc. Natl. Acad. Sci. USA*, **86**, 2708.

30. Thorjmann, T., Berthon, B., Claret, M., and Combettes, L. (1997). *EMBO J.*, **16**, 5398.
31. Ballermann, B. J., Dardik, A., Eng, E., and Liu, A. (1998). *Kidney Int. Suppl.*, **67**, S100.
32. Hamilton, R. L., Berry, M. N., Williams, M. C., and Severinghaus, E. M. (1974). *J. Lipid Res.*, **15**, *182*.
33. Exton, J. H. (1975). In *Methods in enzymology* (ed. J. G. Hardmann and B. W. O'Malley), Vol. XXXIX, p. 25. Academic Press, London.
34. Tran-Thi, T. A., Häussinger, D., Gyufko, K., and Decker, K. (1988). *Biol. Chem. Hoppe-Seyler*, **369**, 65.
35. Häussinger, D. (1989). *J. Hepatol.*, **8**, 259.
36. Häussinger, D. (1996). *Biochem. J.*, **313**, 697.
37. Thomas, A. P. and Delaville, F. (1991). In *Cellular calcium: a practical approach* (ed. J. G. McCormack and P. H. Cobbold), p. 1. IRL Press, Oxford.
38. Morgan, A. J. and Thomas, A. P. (1999). *Methods Mol. Biol.*, **114**, 93.
39. Patel, S., Robb-Gaspers, L. D., and Thomas, A. P. (1999). In *Methods in calcium signaling* (ed. J. W. Putney Jr.), p. 343. CRC Press LLC, Boca Raton, Fl.
40. Svoboda, K., Denk, W., Kleinfeld, D., and Tank, D. W. (1997). *Nature*, **385**, 161.
41. Watanabe, N., Tsukada, N., Smith, C. R., and Phillips, M. J. (1991). *J. Cell Biol.*, **113**, 1069.
42. Gaebler, O. H. (1945). *Am. J. Clin. Pathol.*, **15**, 452.
43. Hanzon, V. (1952). *Acta Physiol. Scand.*, **28**, 4.
44. Carrington, W. A., Lynch, R. M., Moore, E. D. W., Isenberg, G., Fogarty, K. E., and Fay, F. S. (1995). *Science*, **268**, 1483.
45. Agard, D. (1984). *Annu. Rev. Biophys. Bioeng.*, **13**, 191.
46. Monck, J. R., Oberhauser, A. F., Keating, T. J., and Fernandez, J. M. (1992). *J. Cell Biol.*, **116**, 745.
47. Castleman, K. R. (1979). In *Digital image processing* (ed. A. V. Oppenheim), p. 347. Prentice-Hall Inc., Englewood Cliffs, NJ.
48. Hajnóczky, G., Robb-Gaspers, L. D., Seitz, M. B., and Thomas, A. P. (1995). *Cell*, **82**, 415.
49. Robb-Gaspers, L. D., Burnett, P., Rutter, G. A., Denton, R. M., Rizzuto, R., and Thomas, A. P. (1998). *EMBO J.*, **17**, 4987.
50. Hirose, K., Kadowaki, S., Tanabe, M., Takeshima, H., and Iino, M. (1999). *Science*, **284**, 1527.
51. Miyawaki, A., Llopis, J., Heim, R., McCaffery, J. M., Adams, J. A., Ikura, M., *et al.* (1997). *Nature*, **388**, 882.

Chapter 10

Pharmacological studies of new calcium release mechanisms

Antony Galione, Justyn Thomas, and Grant Churchill
Department of Pharmacology, University of Oxford, Mansfield Road,
Oxford OX1 3QT, UK.

1 Introduction

The purpose of this chapter is to describe some of the methods that we use in our laboratory to study cADPR- and NAADP-mediated calcium signalling. Although both of these molecules have been shown to regulate calcium mobilization in mammalian cells, they were discovered as potent calcium mobilizing molecules in sea urchin egg preparations, and this system remains the most powerful in which to study their calcium releasing mechanisms, as well as their metabolism. We therefore describe in detail the major techniques developed in this system, which may be directly extrapolated to other preparations.

2 Preparation of sea urchin egg homogenates and microsomes

The sea urchin egg has been used as a model system in cell biology for many years. For Ca^{2+} signalling studies, its advantages are its large size (100 μm) facilitating microinjection studies, and the production by each animal of several millilitres of packed eggs, all of one cell type, allowing complementary biochemical studies. In addition, microsomal membranes derived from sea urchin eggs contain robust Ca^{2+} release channels, making them the most reliable system for directly studying Ca^{2+} release from intracellular stores.

2.1 Collection of sea urchin eggs

Unfertilized *Lytechinus pictus* sea urchin eggs can be collected during the gravid season between the months of May and September. These eggs are most commonly used on account of their suitability for microinjection and imaging studies. Sea urchins survive shipping if correctly packaged in oxygenated sealed bags of sea water interdispersed with bags of ice. A useful source of *L. pictus* is Marinus Inc., Long Beach CA, USA, and urchins shipped as described above, even survive intercontinental journeys. These urchins survive well in marine aquaria at their natural temperature of 17 °C, and can be kept for many months. Eggs can be obtained by stimulating ovulation of female *L. pictus* urchins by two intracoelomic injections of 0.5 M KCl solution (up to 0.5 ml total volume). After ovulation, urchins can be replaced in the marine tanks and ovulated again after a period of a few weeks. Eggs are collected in artificial sea water (ASW), consisting of: 435 mM NaCl, 40 mM $MgCl_2$, 15 mM $MgSO_4$, 11 mM $CaCl_2$, 10 mM KCl, 2.5 mM $NaHCO_3$, and 1 mM EDTA, adjusted to pH 8.0.

2.2 Sea urchin egg homogenates

Homogenates (50%, v/v) of unfertilized *L. pictus* eggs are readily prepared in a similar manner to that described previously (1).

Protocol 1

Preparation of sea urchin egg homogenates

Equipment and reagents

- 85 μm Nitex mesh (Plastok Associates Ltd., Merseyside, UK)
- Intracellular medium (IM): 250 mM potassium gluconate, 250 mM *N*-methyl glucamine, 20 mM Hepes pH 7.2, 1 mM $MgCl_2$ adjusted to pH 7.2 with glacial acetic acid
- Microcentrifuge
- ASW (see text)
- Homogenization buffer: IM plus 2 mM ATP, 20 U/ml creatine phosphokinase, 20 mM phosphocreatine, 50 μg/ml leupeptin, 20 μg/ml aprotinin, 100 μg/ml SBTI (soya bean trypsin inhibitor)

Method

1. Eggs are dejellied in ASW by filtering through 85 μm Nitex mesh.
2. Dejellied eggs are immediately washed by centrifugation at 800 *g* at 10 °C in approximately ten times their volume of ASW. A hand-held centrifuge is suitable and often more convenient.
3. Eggs are then washed twice in EGTA–Ca^{2+}-free ASW, and twice in Ca^{2+}-free ASW (each comprising 470 mM NaCl, 27 mM $MgCl_2$, 28 mM $MgSO_4$, 10 mM KCl, 2.5 mM $NaHCO_3$ pH 8.0, plus 1 mM EGTA for the first two washes).
4. Finally, eggs are washed with ice-cold 'intracellular medium' (IM).
5. Eggs are then homogenized in an equal volume of homogenization buffer, using a Dounce glass tissue homogenizer, size 'A' pestle. Six gentle strokes is often sufficient.
6. Cortical granules are removed and discarded by pelleting at 13 000 g in a microcentrifuge, at 4 °C for 10 sec. Supernatant is rapidly decanted on completion of the microcentrifuge spin.
7. Homogenates are aliquoted (1 ml) and stored at −70 °C, until used for Ca^{2+}-release studies. We have routinely stored extracts for a number of years without noticeable impairment.

2.3 Fractionation of sea urchin egg homogenate by Percoll density gradient

A Percoll density gradient was used as an established means of obtaining a band of Ca^{2+}-sequestering and releasing microsomes that have been shown to be predominantly endoplasmic reticulum derived (1). In addition it was used to fractionate the sea urchin egg homogenate according to density. A 25% Percoll buffer was prepared for the fractionation, containing an intracellular medium with potassium acetate substituted in place of potassium gluconate, as this provided a more suitable density base for separation of subcellular components.

Protocol 2

Microsome preparation

Equipment and reagents

- Centrifuge and rotor, e.g. 50Ti Beckman rotor
- Syringe and needle
- *Lytechinus pictus* egg homogenate
- Percoll fractionation medium
- Plastic, thin-walled 50 Ti centrifuge tubes

Method

1. *Lytechinus pictus* egg homogenate (1 ml of 50%, v/v) containing an ATP-regenerating system is layered onto the Percoll fractionation medium (9 ml). The final fractionation buffer therefore contained: 250 mM *N*-methyl glucamine, 250 mM potassium acetate, 1 mM $MgCl_2$, 20 mM Hepes; plus an ATP-regenerating system comprising: 500 μM ATP, 2 U/ml creatine phosphokinase, 4 mM phosphocreatine, and 25% (v/v) Percoll pH 7.2.
2. The tubes are centrifuged at 27 000 g, for 30 min at 10 °C in a 50Ti Beckman rotor. This temperature allows microsomal loading by calcium, and results in a tighter microsomal band (*Figure 1*).
3. The resulting fractions are either collected by gently removing 1 ml layers from the top, or the microsomal band was removed by puncturing the vessel wall (*Figure 1*) using a syringe and needle. The top 'cytosolic' fraction contains enzymes for synthesis of cADPR, as well as cofactors for release mechanisms such as calmodulin.
4. Fractions are stored at −70 °C until required.

3 Studying activation of calcium release channels

3.1 Fluorimetry of Ca^{2+} release from egg homogenates

Ca^{2+} loading of intracellular stores was achieved by incubating homogenates for 3 h at 17 °C in IM containing the additions for ATP-regenerating system, mitochondrial inhibitors (1 μg/ml oligomycin and antimycin, 1 mM sodium azide), and protease inhibitors as outlined above. Use of the regenerating system allows the resynthesis of ATP and keeps ATP levels reasonably constant.

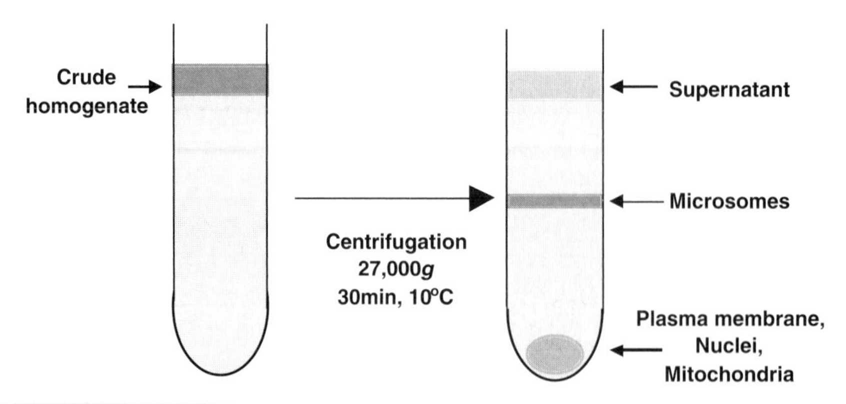

Figure 1 Percoll density gradient fractionation of sea urchin egg homogenates. 1 ml of 50% (v/v) crude sea urchin egg homogenate was layered onto 9 ml of a 25% Percoll solution (containing acetate–IM). The gradient was formed and fractionation at 27 000 *g* for 30 min at 10 °C. The supernatant, microsomes, and other fractions form as indicated. The microsomal band can be removed by puncturing the wall of the plastic tube with a syringe and collected in approximately 500 μl.

Protocol 3

Fluorimetric Ca^{2+} release from sea urchin egg homogenates

Equipment and reagents

- Fluorimeter, e.g. Perkin Elmer LS-50B/PTI
- IM buffer (see *Protocol 1*)
- ATP-regenerating system (see *Protocol 2*)
- Protease inhibitors (see *Protocol 1*)
- Mitochondrial inhibitors: 1 μg/ml oligomycin and antimycin, 1 mM sodium azide
- Fluorescence indicator, e.g. fluo-3

Method

1. Over 3 h the homogenate (50%) is sequentially diluted. After the first dilution (1:1, v/v), the homogenate is incubated for 1–2 h. Then every half-an-hour more 1:1 dilutions are carried out. This results in a dilution from 50% homogenate (v/v), to a final dilution of 2.5% (v/v) in the IM buffer. This method for preparation has been found empirically to achieve better loading of vesicles and highly reproducible lower steady state resting Ca^{2+} levels than by a one-step dilution.
2. In the final 30 min 3 μM fluo-3 is added as the Ca^{2+} fluorescence indicator.
3. Free Ca^{2+} is measured at 17 °C using 500 μl of homogenate in a Perkin Elmer LS-50B/PTI fluorimeter at excitation and emission wavelengths of 490–505 nm and 526–535 nm, respectively.
4. Additions of drugs are made to the homogenate in 5 μl volumes, and changes in relative fluorescence units (ΔRFU) are calibrated to known Ca^{2+} additions (*Figure 2*) expressed as nmol of calcium.

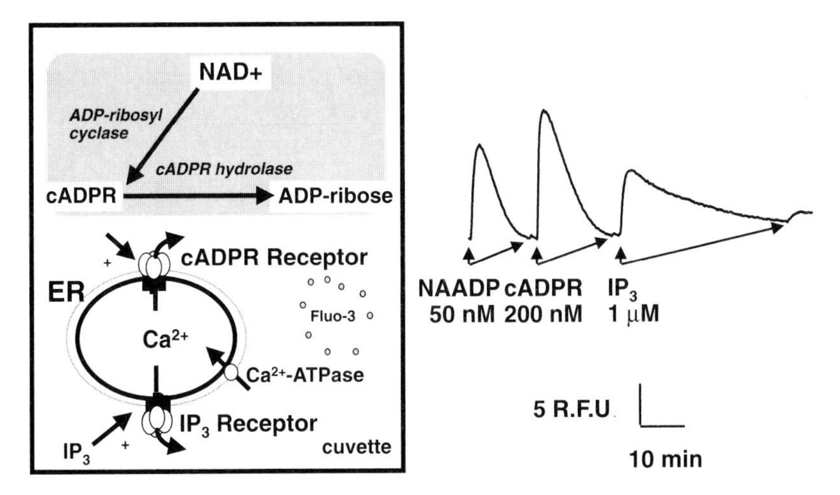

Figure 2 Fluorimetric Ca^{2+} release from sea urchin egg homogenates. Fluorimetry of egg homogenates using fluorescent Ca^{2+} indicators have allowed a detailed analysis of the metabolism and mechanism of cADPR-induced Ca^{2+} release. Egg homogenates/microsomes are incubated with ATP to allow loading of Ca^{2+} stores via Ca^{2+}-ATPases. Addition of Ca^{2+} mobilizing agents elicit a translocation of Ca^{2+} from stores into the extravesicular medium which is detected by changes in fluorescence of indicators (depicted as the trace in the right hand part of the figure) added to the extravesicular medium in the cuvette. Traces are shown of fluo-3 fluorescence (ex. 490 nm/em. 535 nm) measuring extravesicular calcium during cuvette-based fluorimetry of sea urchin egg homogenates. The homogenates sequester calcium in the presence of an ATP-regenerating system and release calcium in response to calcium mobilizing agents. At least three distinct calcium release mechanisms are present, activated by NAADP, cADPR, or IP_3. A feature of calcium release systems in this preparation is that each mechanism undergoes a homologous desensitization as indicated by the failure of a second application of each calcium mobilizing agent to produce a second calcium release even after full resequestration of calcium.

3.2 Testing microsomal fractions for Ca^{2+} releasing properties

Percoll fractions (approx. 50%, v/v) were defrosted and immediately used for fluorimetry. Ca^{2+} sequestration was achieved by adding 50–100 μl of the fraction with 400–450 μl IM (250 mM *N*-methyl glucamine, 250 mM potassium gluconate, 1 mM $MgCl_2$, and 20 mM Hepes pH 7.2). The increased density of microsomal membranes in these fractions compared to the whole homogenate, means that long pre-incubations to allow Ca^{2+} sequestration are not necessary, and the microsomes will sequester Ca^{2+} within 5–10 minutes. In addition, an ATP-regenerating system was provided (1 mM ATP, 10 U/ml creatine phosphokinase, and 10 mM phosphocreatine) plus mitochondrial inhibitors (1 μg/ml oligomycin and antimycin, 1 mM sodium azide) and protease inhibitors (as in previous section) were included to prevent mitochondrial Ca^{2+} release and protein degradation, respectively. Ca^{2+} sequestration and release to various agents was detected fluorimetrically, using 3 μM fluo-3 as a Ca^{2+} indicator.

3.3 Pharmacology of cADPR signalling

The pharmacology of cADPR-induced Ca^{2+} release has been most extensively investigated in the sea urchin egg system. cADPR was found to activate and modulate Ca^{2+}-induced Ca^{2+} release (CICR) via a ryanodine-sensitive Ca^{2+} release channel (2, 3) independent of the activation of inositol trisphosphate receptors. The first indication that the β-NAD^+ metabolite cADPR mobilized Ca^{2+} via a mechanism distinct from IP_3-gated Ca^{2+} release channel, came from studies in sea urchin egg homogenates whereby Ca^{2+} release by β-NAD^+ or IP_3 rendered the Ca^{2+} release mechanism refractory to a second application of the same agent but not to the other.

Pharmacological analysis confirmed that cADPR did not activate IP_3 receptors (IP_3Rs), since the competitive IP_3R antagonist heparin blocked IP_3-induced Ca^{2+} release but not that triggered by cADPR. The identity of the Ca^{2+} release mechanism activated by cADPR came from studies demonstrating the presence of a ryanodine receptor (RyR) mechanism in the sea urchin egg. Two pharmacological activators of RyRs, caffeine and ryanodine, induced Ca^{2+} release in both intact eggs (2, 3) and in sea urchin egg homogenates, but precluded a subsequent release of Ca^{2+} by cADPR but not IP_3. Sea urchin egg microsomes that had discharged their Ca^{2+} contents in response to cADPR were rendered insensitive to caffeine and ryanodine also but not to IP_3. Furthermore, blockers of RyRs such as ruthenium red and procaine selectively inhibited cADPR-induced Ca^{2+} release. These later experiments led to the hypothesis that cADPR may be an important regulator of RyRs, and hence CICR, an important property of RyRs whereby a small rise in cytoplasmic Ca^{2+} triggered a larger Ca^{2+} release from internal stores by activating RyRs (2). Consistent with this hypothesis was the finding that in sea urchin egg homogenates and in intact cells, cADPR potentiated Ca^{2+} release by Ca^{2+} and Sr^{2+}, and caffeine (3), and that cADPR-induced Ca^{2+} release was inhibited by magnesium ions (4). Sr^{2+} although less potent at triggering CICR in this system, has the advantage that at triggering concentrations (1–50 μM) it does not appreciably affect fluo-3 fluorescence. Thus release of calcium (CICR) is clearly seen without masking by trigger calcium.

Since the RyR hypothesis for cADPR action was formulated, cADPR has been shown to mobilize Ca^{2+} in many cell types. Where the pharmacology has been examined the majority of studies have shown that the pharmacology of release display similarities to RyRs (5). The development of highly selective cADPR analogues, including caged compounds (6), competitive antagonists (7), metabolically-resistant agonists and antagonists (8), and membrane permeant analogues (9) have led to important advances in cADPR research. These compounds have therefore been of great benefit in probing the fundamental mechanisms governing cADPR-mediated signalling mechanisms and pathways.

As mentioned above, the pharmacology of cADPR-mediated Ca^{2+} release has been most extensively studied in sea urchin egg homogenate and microsomes. However, the techniques used can and have been applied to microsomal systems from mammalian cells as well as permeabilized cells (10, 11). The fluorimetric technique for measurement of Ca^{2+} has been used widely, but in mammalian systems ^{45}Ca efflux studies have also been employed. The choice of intracellular medium for such studies may be important. Ca^{2+} release in the sea urchin homogenate system for example, is more sensitive to releasing agents in the *N*-methyl glucamine/potassium gluconate medium than in a KCl medium, while replacement of gluconate with mannitol, improves the sensitivity of the IP_3 response (12). In trying to optimize Ca^{2+} release in broken cell systems, it is clear that different media should be tested.

With the fluorimetry approach to measuring Ca^{2+} release, there are several potential pitfalls and artefacts that should be avoided, and extensive controls should be performed to avoid the drawing of invalid conclusions. Ca^{2+} contamination of cADPR can be a problem, and care should be taken in dissolving samples in the purest water available. In addition, low concentrations of EGTA can be added to the samples, but this should not be so high as to reduce or abolish the detection of the Ca^{2+} release response. A few beads of Chelex or molecular calcium sponge can be added to aliquots of cADPR containing solutions, which can be then spun down and the supernatant decanted for use. However, since Chelex can also sequester other molecules, steps such as verifying the continued presence of cADPR or other molecules in the treated solutions by HPLC could be carried out (see Section 4.4).

In the sea urchin egg homogenate, a second application of a maximal cADPR concentration does not induce further Ca^{2+} release due to a poorly understood desensitization process. This has also been observed in other but not all cell preparations. This provides a useful check for Ca^{2+} contamination, since a second fluorescent peak may be indicative of such contamination.

Another useful check is to heat inactivate solutions containing cADPR. Heating the solution to 90 °C for several hours non-enzymatically converts cADPR to ADPR which does not mobilize Ca^{2+} from intracellular stores. However, ADPR may have other cellular effects, for example the potent gating of a cation current in intact ascidian oocytes (13).

Verification of an authentic Ca^{2+} release by cADPR has come from the use of pharmacological modulators of the Ca^{2+} release channel. Competitive antagonists of cADPR-induced Ca^{2+} release have been synthesized (7). 8-NH_2-cADPR, while blocking cADPR-induced Ca^{2+} release, does not impair the ability of caffeine or ryanodine to elicit Ca^{2+} release from sea urchin egg homogenates. These data may point to the complex nature of the regulation of RyRs which are subject to allosteric regulation by small molecules as well as by the many different proteins with which they have been shown to interact (14). This notion is reinforced by recent investigations that showed a requirement for calmodulin for cADPR-induced Ca^{2+} release from purified sea urchin egg microsomes. This may provide one explanation for the failure of some investigators to observe cADPR-mediated Ca^{2+} release in some broken cell systems.

In addition, it has been found in many systems that selective blockers of RyR-mediated Ca^{2+} release, reduce or inhibit that induced by cADPR. Ryanodine itself, ruthenium red, procaine (2), and millimolar concentrations of Mg^{2+} (4)

have all been used to inhibit cADPR actions, whereas caffeine, at low millimolar concentrations, is able to sensitize Ca^{2+} evoked by cADPR (3). Care should be taken that possible interference with Ca^{2+} indicator fluorescence should not obscure results. Controls should be performed to test for any interactions with the dyes and if necessary correct ions for any interference made (15). Different Ca^{2+} indicators can also be tried.

3.4 Studying NAADP-induced Ca^{2+} release

As with the study of the role of cADPR in Ca^{2+} signalling, the sea urchin egg has proven to be an invaluable model to determine the primary characteristics of the novel NAADP Ca^{2+} release mechanism (16). In addition to the two species of sea urchins that live off the coasts of the USA: *L. pictus* and *Strongylocentrotus purpuratus*, this novel signalling molecule has also been shown to be present in *Paracentrotus lividus*, a common sea urchin in the Mediterranean sea, and in *Psammechinus miliaris*, a sea urchin that lives off the coast of Britain (17). Therefore, any of these species, and probably other sea urchins as well, are suited to studies to investigate NAADP-mediated signalling. The methods to explore Ca^{2+} signals elicited by NAADP in sea urchin eggs are similar to those described in the previous sections for cADPR with some modifications that will be explored in detail.

3.5 Desensitization of NAADP-induced calcium release

NAADP-mediated Ca^{2+} signalling in the sea urchin egg has been shown to possess a feature that makes it unique: pre-treatment of homogenates with minute concentrations of NAADP, that *per se* do not elicit any appreciable Ca^{2+} release can completely abolish further Ca^{2+} release induced by concentrations of NAADP that elicit maximal Ca^{2+} release (18, 19) (*Figure 3*).

All four species of sea urchins named above exhibit this paradoxical inactivation property. This feature, which is not a property of either cADPR- or IP_3-induced Ca^{2+} release, can influence all protocols dealing with NAADP and therefore some experimental strategies must be designed to avoid artefacts. NAADP is very potent in releasing Ca^{2+} and even more potent in inactivating the receptor, with EC_{50}s and IC_{50}s of approximately 32 nM and 200 pM, respectively. Therefore, small contaminations present on used tips or plasticware, can alter the amount of Ca^{2+} release. Furthermore, the rate of delivery of NAADP to the homogenate can also greatly affect the extent of release. If NAADP is presented to the receptor in a slow manner (slow release during pipetting or in intact cells diffusion through a patch pipette), the response will not be as high as when NAADP is presented to the receptor in a rapid manner (therefore rapid mixing and consistent pipetting are crucial to have reproducible results). The inactivation phenomenon can also greatly affect the study of pharmacological agents that act on the NAADP receptor, particularly when structural analogues are examined. For example, thio-NADP, which had been used as an antagonist of putative NAADP receptors, has been more recently shown to been shown to be contaminated with NAADP itself (20), which may explain the antagonism.

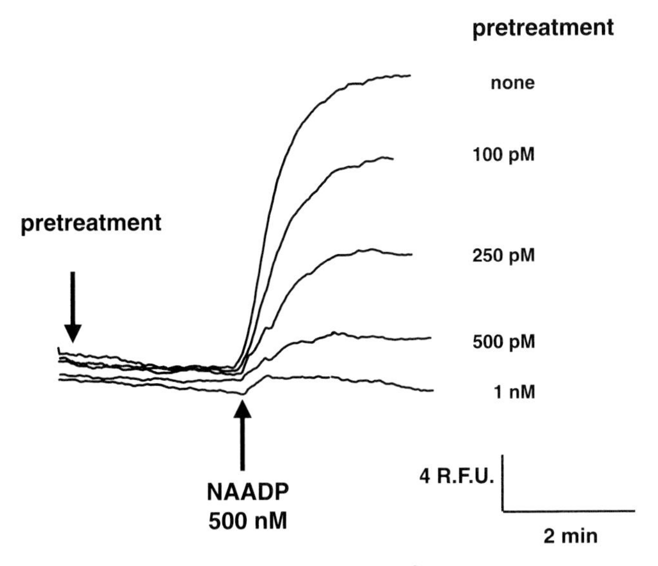

Figure 3 Inactivation of NAADP-induced Ca^{2+} release by NAADP. Representative fluorimetric traces of Ca^{2+} release by 500 nM NAADP (top trace) and of homologous desensitization by sub-threshold concentrations of NAADP in 2.5% sea urchin homogenate. Homogenates were pre-treated for 3 min with the different concentrations of NAADP. The homogenate was then challenged with a maximum concentration of NAADP (500 nM). Ca^{2+} release is expressed as relative fluorescence units (RFU).

NAADP has been shown to release Ca^{2+} from pools that are distinct from those emptied by IP_3 and cADPR. It has been shown that emptying of the stores with thapsigargin abolishes Ca^{2+} release elicited by cADPR or IP_3, but does not affect that evoked by NAADP (21). This has been strengthened by the observation that NAADP-responsive microsomes migrate differently from IP_3 and cADPR-responsive Ca^{2+} pools, which migrate in a single band together with endoplasmic reticulum markers (1). Therefore, if attempting to enrich by fractionation preparations of NAADP-sensitive pools, it is important not to assume that this pool will migrate with endoplasmic reticulum standards nor with pools responsive to IP_3 or cADPR.

To verify that a response is indeed due to NAADP, pharmacological analysis is recommended. NAADP responses in the sea urchin homogenate have been shown to be unaffected by pre-treatment with concentrations of heparin and 8-NH_2-cADPR that fully block IP_3 and ryanodine receptors. Furthermore, a number of agents have been shown to block NAADP receptors selectively over the ones responsive to IP_3 and ryanodine receptors . In particular, L-type Ca^{2+} channel blockers and K^+ channel blockers appear to discriminate between intracellular receptors (22). This is also true for T-type Ca^{2+} channel blockers, such as amiloride (100 μM), flunarizine (100 μM), and pimozide (25 μM) (Genazzani and Galione, unpublished data). Unfortunately, the concentrations of these drugs needed to block the NAADP response appear to be at least tenfold higher than those required to block their main targets, making these compounds unsuited to be used in intact cells. Calmodulin antagonists, such W7 and J8, also abolish NAADP-induced Ca^{2+} release but at similar concentrations to those needed to block cADPR- and IP_3-induced Ca^{2+} release. TMB-8 has also been reported to block NAADP-induced Ca^{2+} release in sea urchin eggs (23), although also in this case the concentrations required (500 μM) are unlikely to dissect between the different mechanisms. As mentioned above, the activity of the only antagonist reported in the literature, thio-$NADP^+$, has been shown to be due to inactivation of the release mechanism by contaminating NAADP. An initial structure–activity study has also been performed, but while it has determined important structural requirements for NAADP analogues to be active, it failed to find specific antagonists of this mechanism (24).

Important differences between NAADP-induced Ca^{2+} release and the other two mechanisms are depicted by the different sensitivities to pH, Ca^{2+}, and monovalent cations. NAADP-induced Ca^{2+} release, unlike the cADPR- and IP_3-sensitive mechanisms, is unaffected by cytosolic Ca^{2+} concentrations, pH (between 6.7 and 8.0), and the presence of different monovalent cations in the experimental buffer, and thus the mechanism is not subject to dual regulation by messengers and Ca^{2+}.

4 Measuring production and degradation of second messengers that trigger calcium signals

4.1 Enzymatic synthesis and degradation of cADPR

The synthesis of cADPR is catalysed by an enzymatic cyclization reaction, in which nicotinamide is released from the substrate β-NAD^+, as outlined in *Figure 4*. The finding that sea urchin egg homogenates contain enzymatic activities to synthesize cADPR from β-NAD^+ has been extended to many other animal tissues. The synthetic enzymes, known as ADP-ribosyl cyclases, are widespread in tissues (25). The first molecularly characterized form was found in *Aplysia* ovotestis, which on account of its high cyclase activity, monofunctional activity, and loose substrate specificity has been used for the chemo-enzymatic synthesis of many cADPR analogues (26). Such analogues have subsequently been used to probe cADPR signalling. In addition, ADP-ribosyl cyclase activities have been reported in cytoplasmic compartments in sea urchin eggs and to be associated with cardiac mitochondria and sarcoplasmic reticulum.

Enzymes that hydrolyse cADPR to ADP-ribose (cADPR hydrolases) are equally widespread (27). An intriguing finding is that in many tissues, both cyclase and hydrolase activities are expressed on the same bifunctional polypeptide. The most well characterized example of such a protein is the cell surface antigen CD38 (28).

4.2 NAADP metabolism

ADP-ribosyl cyclases such as CD38 can catalyse two types of reaction, cyclization and base exchange (*Figure 4*), and use β-NAD^+ and β-$NADP^+$ as alternative sub-

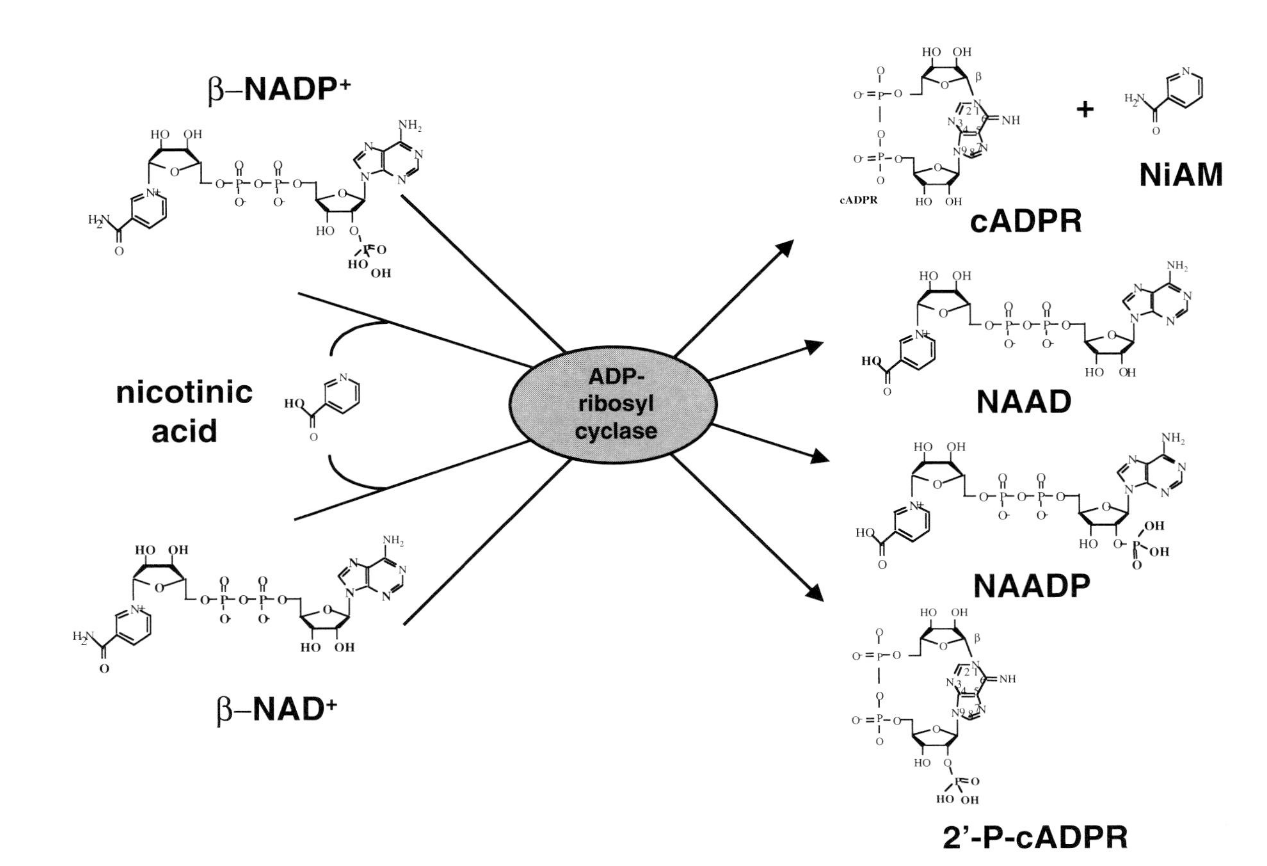

Figure 4 Reactions catalysed by ADP-ribosyl cyclases. β-NAD$^+$ and β-NADP$^+$ can be cyclized to cADPR or 2-P'-cADPR as the respective products. In the presence of nicotinic acid, a base-exchange reaction is catalysed, with β-NAD$^+$ converted to NAAD and NAADP formed from β-NADP$^+$. In all cases nicotinamide (NiAM) is released.

strates. In the base-exchange mode, nicotinamide is exchanged for nicotinic acid, yielding nicotinic acid adenine dinucleotide (NAAD) and nicotinic acid adenine dinucleotide phosphate (NAADP), depending on whether β-NAD^+ or β-$NADP^+$ are used as substrates, respectively. Enzymatic synthesis and degradation of NAADP has been detected in rat brain tissues, but has remained uncharacterized in most other tissues. If β-$NADP^+$ is cyclized, the product is 2-phospho-cADPR (cADPRP). Cyclic ADP-ribose phosphate appears more potent than cADPR in activating RyRs in brain microsomes (38) and has been shown to release Ca^{2+} in Jurkat cells but is ineffective in sea urchin egg homogenates. This raises the intriguing possibility that subtypes of cADPR receptor exist.

cADPR production can be monitored by fluorimetry by incubating the tissue homogenates with adequate concentrations of NAD^+. As described above, NAD^+ will elicit Ca^{2+} release after a delay (latency). Both the latency and the peak Ca^{2+} release are good parameters for the estimation of the rate of cADPR production. Due to the inactivating property of the NAADP-sensitive mechanism, this 'real time' protocol cannot be used with this messenger. Furthermore, it has been shown that NAADP can be produced by cells from $NADP^+$ only in the presence of nicotinic acid, by a base-exchange reaction (nicotinamide is exchanged for nicotinic acid). The most popular way to measure NAADP production therefore is to use a bioassay similar to the one described for cADPR. The homogenate is incubated with $NADP^+$ and nicotinic acid and an aliquot of this incubation is assayed in the fluorimeter in a fresh homogenate preparation. The amount of Ca^{2+} release by this aliquot is then compared to a dose–response curve constructed with authentic NAADP to determine the amount of NAADP that has been produced. To render the assay more sensitive, a concentration–response of NAADP inactivation can also be performed and the amount of residual release to a maximal NAADP concentration after 3 min can be measured and related to standards. This assay for NAADP levels has a threshold sensitivity of 3 pmoles compared to a threshold sensitivity to cADPR levels of approximately 500 pmoles. In addition to the measurement of NAADP production in sea urchin eggs, the same procedure can be used to quantify the capacity of mammalian and plant cells to produce NAADP (39). In a similar manner, tissues can be incubated with authentic NAADP and degradation of this molecule can be followed by the bioassay. In our experience, concentration of 1–10 mM nicotinic acid and 100–500 μM $NADP^+$ are necessary to have detectable NAADP production with the bioassay. As mentioned above, it has been shown that the same enzymes responsible for cADPR production (e.g. *Aplysia* cyclase, CD38) are also responsible for the base-exchange reaction that leads to NAADP. In line with these data, we find that in mammalian tissues, millimolar concentrations of nicotinamide and micromolar concentrations of cibacron blue, known to inhibit cADPR production, also inhibit NAADP formation (Genazzani and Galione, unpublished observations). Although HPLC separation and quantification of NAADP on AG-1 HPLC columns has been described, in our hands this method has not proved to be of enough sensitivity to determine levels with ease.

4.3 Detection of cADPR and NAADP metabolism by discontinuous bioassay

The ability of a cell homogenate system to synthesize or degrade cADPR can be readily determined using the sea urchin egg homogenate as a bioassay.

Protocol 4

Measuring cADPR and NAADP metabolism

Reagents

- β-NAD$^+$ or cADPR
- β-NADP$^+$ and nicotinic acid, or NAADP
- IM buffer (see *Protocol 1*)
- Sucrose–Hepes buffer: 250 mM sucrose, 20 mM Hepes pH 7.2

Method

1. Cell homogenates are incubated with β-NAD$^+$ (usually 2.5 mM), or cADPR (5 μM), respectively. The concentrations stated are final in the incubation mixture, and therefore stock solutions should be approximately 100 × higher.
2. The synthesis and degradation of NAADP is determined by incubating with β-NADP$^+$ (250 μM) and nicotinic acid (7 mM), or NAADP (5 μM) itself, respectively.
3. Incubations are maintained at 17 °C in IM buffer for sea urchin egg homogenates or at 37 °C in sucrose–Hepes buffer for mammalian cell homogenates.
4. Synthesis or degradation of the respective Ca^{2+} releasing agents is monitored over time, where 5 μl of the reaction mixture was withdrawn and assayed immediately, using the sea urchin egg homogenate as a bioassay.
5. The specificity of release is confirmed by desensitization to standard chemicals, or by blockade using specific antagonists. For substrate incubations an increase in the appropriate Ca^{2+} releasing activities with time indicated the presence of synthetic enzymes in the homogenate tested.
6. The presence of degradative enzymes, e.g. cADPR hydrolases is predicted from a decrease in the Ca^{2+} release from discontinuous monitoring over time, for incubations with either cADPR or NAADP.
7. We have found that inclusion of the ADP-ribosyl cyclase inhibitor nicotinamide (10 mM) (29) in the test homogenate prevents any possible conversion of unreacted NAD to cADPR by egg homogenate enzymes.

4.4 NGD assay: continuous assay for GDP-ribosyl cyclase activity

The ADP-ribosyl cyclase activity of a given cell homogenate can also be detected by a continuous assay, using NGD$^+$ instead of β-NAD$^+$ as substrate to yield the fluorescent product cGDPR (30).

Protocol 5

cGDPR production

Equipment and reagents

- Perkin Elmer LS50-B fluorimeter
- See *Protocol 4*
- 250 μM NGD^+

Method

1. Reaction mixtures contained 50–100 μl of the cell homogenate being tested, and made up to a final volume of 500 μl using sucrose–Hepes or IM buffer pH 7.2.
2. The production of the fluorescent product, cyclic GDP-ribose, can be measured using a Perkin Elmer LS50-B fluorimeter, with excitation and emission wavelengths of 300 nm and 410 nm (*Figure 5*), respectively.
3. When a basal fluorescence is recorded, the reaction is initiated with the addition of 250 μM NGD^+. The assay is not as sensitive as the bioassay since for example in sea urchin egg homogenates GDP-ribosyl cyclase activity was not able to be detected using this assay, as either the production was too low, or the homogenate too turbid to measure changes in fluorescence.
4. In addition, other cADPR analogues can be used to measure cADPR hydrolase activities. For example, if the fluorescent analogue cIDPR is used as a substrate, then hydrolysis is indicated by a fall in fluorescence, as it is a good substrate for cADPR hydrolases (30).

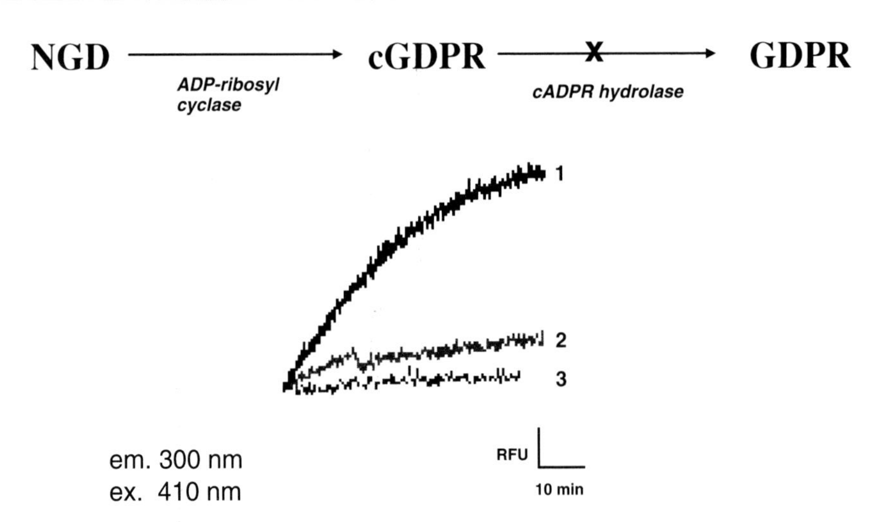

Figure 5 cGDPR production in rabbit vascular smooth muscle. Muscle homogenates (400 μg/ml) were monitored continuously by fluorimetry in a cuvette maintained at 37 °C, using excitation and emission wavelengths of 300 nm and 410 nm respectively. The substrate NGD^+ was added and the change in fluorescence due to its cyclization to cGDPR monitored over time (1). Similar experiments were performed in the presence of nicotinamide which blocks cyclization (2) or in the absence of NGD^+ (3). Data taken from ref. 42.

4.5 HPLC

Reaction products formed as a result of NAD and NADP metabolism can also be verified by HPLC protocols. In addition, when ascertaining synthesis of either cADPR or NAADP, it is important to check that their respective precursors NAD and NADP are not contaminated with these molecules. If so, they can be purified. It is interesting to remember that the Ca^{2+} mobilizing activity of NAADP was discovered as a contaminant of commercially available β-$NADP^{+}$. HPLC protocols have been used in the measurement of cADPR levels in cells as described briefly in Section 4.6 and in more detail in *Protocol 7*, outlining the synthesis and isolation of [^{32}P]cADPR.

4.6 Measurement of cADPR levels in cells and tissues

A number of different techniques have been used to assess levels of cADPR in cells and tissues. These include microsomal bioassays using sea urchin egg homogenates (31), thin-layer chromatography (32, 33), radioimmunoassay (34), radioreceptor assay (35), and HPLC (36). The first clue as to how ADP-ribosyl cyclase activities can be regulated came from studies in the sea urchin egg, which examined the mechanism of cGMP-induced Ca^{2+} release in these cells. When microinjected into sea urchin eggs, cGMP induces a large, prolonged Ca^{2+} transient, after a considerable latency, reminiscent of the Ca^{2+} transient seen in the fertilizing egg. Pharmacological studies indicated that RyRs were likely to be involved since this effect of cGMP is blocked by the RyR inhibitor ruthenium red but not by the IP_3R antagonist heparin. Using sea urchin egg homogenates to dissect the cGMP pathway, it was found that cGMP-induced Ca^{2+} release could be reconstituted only if homogenates were supplemented with the cADPR precursor β-NAD^{+}. In this system it was also found that Ca^{2+} release by cGMP was blocked by 8-amino-cADPR, a selective competitive antagonist analogue of cADPR, but not by heparin, suggesting that cADPR mediated the effect, and TLC analysis showed that cGMP stimulated cADPR production from β-NAD^{+} (32). A role for ADP-ribosyl cyclase activity was implicated by the finding that the sea urchin egg ADP-ribosyl cyclase inhibitor, nicotinamide, blocked cGMP-induced Ca^{2+} transients in both sea urchin egg homogenates and in intact cells (29).

Protocol 6

Preparation of samples for analysis of cell and tissue cADPR levels

Equipment and reagents

- Sonicator
- Centrifuge
- 0.5 M perchloric acid
- [^{3}H]cADPR
- 2 M $KHCO_3$

Protocol 6 continued

Method

1. Cells and tissues can be treated with agonists and then plunged into ice-cold perchloric acid solution to stop the reaction and precipitate proteins.
2. [^{3}H]cADPR (10 000 c.p.m.) can be added to the samples to determine recovery.
3. The samples are then quickly vortexed or sonicated and centrifuged at 15 000 g for 10 min at 4 °C.
4. Supernatants are then neutralized with 2 M $KHCO_3$, and then centrifuged again at 15 000 g for 10 min at 4 °C. The pellet of precipitated potassium perchlorate is discarded.
5. The supernatant can then be frozen by snap freezing in liquid nitrogen and stored at −70 °C until analysis.
6. Samples can then be analysed for cADPR levels by bioassay (31) (see below), radioreceptor assay (Section 5.1), or HPLC (36).

Using the sea urchin egg homogenate as a bioassay has been very useful in examining cADPR metabolism. The advantage of the bioassay is in its specificity. Cross-desensitization between a sample and authentic cADPR in terms of Ca^{2+} release, or inhibition by 8-NH_2-cADPR, provide good evidence that the Ca^{2+} mobilizing effects can be ascribed as due to cADPR. Standard curves for Ca^{2+} release by known amounts of authentic cADPR can be used to determine amounts of cADPR in tissue samples. A detailed description of this method has been given previously (31). The drawback of the technique is in its sensitivity and its susceptibility to perturbation by other factors in the tissue sample as the level of divalent cations. As mentioned above Ca^{2+} ions potentiate cADPR-induced Ca^{2+} release while magnesium ions inhibits it. Chelation of Ca^{2+} in the samples by adding Chelex beads or magnesium by adding EDTA, offers a solution to these difficulties. The low sensitivity to cADPR means that tissue samples need to be concentrated and in so doing also concentrates molecules or ions that could interfere with the assay. We have found that a tissue substance that is concentrated along with cADPR, interferes with the fluorescence of a number of Ca^{2+} indicators that we routinely use for fluorimetry. One potential way around this is to add an HPLC step to resolve cADPR from other substances, but unfortunately it co-elutes with cADPR. The sensitivity of the assay can be improved by up to tenfold by adding 1 mM caffeine or 50 μM thimerosal to test homogenates. These are threshold concentrations of these agents which potentiate cADPR-induced Ca^{2+} release.

Polyclonal antibodies raised against cADPR have recently been reported (34). These, while having a high affinity for cADPR often cross-react with other metabolites such as NAD which are present in cells at significantly higher concentrations than cADPR. Some are commercially available (Chemicon) but very expensive. An advantage over the bioassay is not only their increased sensitivity, but that binding of cADPR is less susceptible to modulation by other agents. To remove nucleotides which interfere with the assay, a number of enzymes can be added to the tissue samples. NADase from *Neurospora crassa*, nucleotide pyrophosphatase from *C. atrox* venom, alkaline phosphatase from bovine intestinal mucosa, and apyrase from potato have been used to degrade NAD, ATP, and other nucleotides, but do not metabolize cADPR (34). To check the authenticity of the assay, the effects of tissue sample on binding of radiolabelled cADPR to the antibody, can be compared with the effects of the tissue sample incubated with CD38 which has a high cADPR hydrolase activity, and thus should be devoid of cADPR. A procedure for the large scale expression and isolation of a soluble form of CD38 has been reported which is useful for this assay (37). Both [^{32}P]cADPR (see Section 5.1) and the commercially available [^{3}H]cADPR have been used as radioligands, but the former is superior on account of its higher specific activity generating more counts for more accurate determinations.

5 Binding of agonists to calcium release channels

5.1 cADPR binding studies

Synthesis of [^{32}P]cADPR is desirable for identification of cADPR binding sites and may also be used in a radioreceptor assay for determination of endogenous cADPR levels. This compound has a much high specific activity (~ 1000 Ci/mmol) than its tritiated commercially available counterpart.

Protocol 7

Preparation of [^{32}P]cADPR

Equipment and reagents

- HPLC equipment
- Microcentrifuge tubes
- Scintillation counter
- [^{32}P]-β-NAD^{+} (Amersham)
- ADP-ribosyl cyclase
- Tris–HCl pH 7.4
- TFA

Method

1. To prepare [^{32}P]cADPR, the cyclization reaction is carried out for 2 h at room temperature in the vial in which the radiolabelled compound is shipped.
2. The reaction solution contains 250 μCi [^{32}P]-β-NAD^{+}, 100 ng/ml of repurified ADP-ribosyl cyclase, and 5 mM Tris–HCl pH 7.4 in a total volume of 250 μl.
3. Cyclic ADP-ribose is separated from its substrate using an HPLC method (40) that limits the radioactive contamination to the column and some tubing that is dedicated to radioactive work and prevents contamination of the injection port and the absorbance detector. This method also enables all the procedures with radioactive compounds to be performed in the same space a few feet away from the HPLC

Protocol 7 continued

equipment behind Perspex (Plexiglass) shielding. All work with ^{32}P is carried out behind Perspex shielding while wearing gloves and goggles.

4. The reaction mixture is loaded onto a column with a peristaltic pump (flow rate of about 100 μl/min) by inserting the end of peek tubing (attached via a peristaltic tubing adapter; Upchurch) into the reaction vial.
5. The reaction vial is washed with water 1 ml at a time (total volume 5 ml) until the radioactivity, as detected with a Geiger counter, remaining in the reaction vial does not decrease any further and the majority of the radioactivity is at the top of the column. The connection between the peek tubing attached to the column and the peristaltic tubing is then broken, and reattached via a female–female peek union (Upchurch) to a length of peek tubing coming from the outport of the HPLC pump. 5 ml of water is pumped through the column to ensure there are no leaks and then the TFA gradient is started. The bottom of the column has a short piece of peek tubing attached and the effluent is collected as 1 ml fractions in microcentrifuge tubes. Peaks are determined by scintillation counting 1 μl of each fraction in 10 ml water using Cerenkov radiation. β-NAD^+ elutes at 7 min, cADPR at 15 min, and ADPR at 20 min.
6. Fractions containing cADPR (12–16 min) are neutralized with Tris base, with the concentration of Tris base estimated from the amount of TFA in the given fraction, and then stored frozen at −20 °C.

As an example of cADPR binding we present data for binding to whole-egg homogenates from the sea urchin *Lytechinus pictus*. About 40 000 c.p.m. (180 pM) of [^{32}P]cADPR and a series of concentrations of cold cADPR are incubated in 250 μl of 1 mg/ml homogenate in Glu-IM for 30 min on ice. Bound and free radioligand are separated by filtration with a cell harvester (Brandel) onto glass fibre filter paper (Whatman B), that is washed immediately before filtration with ice-cold Glu-IM, and washed twice with 2–4 ml of ice-cold Glu-IM immediately after filtration. The resulting filters are placed into scintillation tubes with 10 ml water and counted in a scintillation counter using Cerenkov radiation. Specific binding was calculated by subtracting total binding from non-specific (10 μM cold cADPR). *Figure 6* shows a homologous competitive binding curve for [^{32}P]cADPR displaced with cold cADPR. The resulting curve yields a K_d of 1.1 nM, which is similar to that previously reported (41) and a B_{max} of 9.2 fmol/mg. This assay has recently be used to determine changes in cADPR levels in hippocampal brain slices using [^3H]cADPR as the radioligand during induction of long-term depression (35).

5.2 The NAADP receptor

Radioligand [^{32}P]NAADP binding. Binding experiments have shown that NAADP binds irreversibly to the receptor, providing a possible explanation for the in-

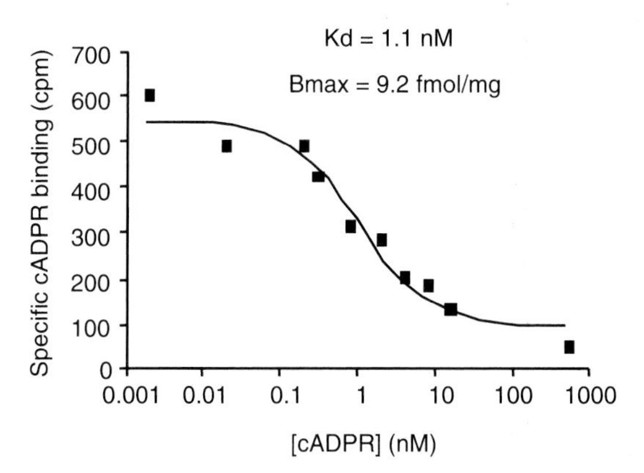

Figure 6 [^{32}P]cADPR binding to sea urchin egg microsomes. Homologous competitive binding curve for [^{32}P]cADPR and sea urchin homogenate. The curve and K_d are based on non-linear curve fitting to a single-site model using a logistic equation (r_2 = 0.96) with the software GraphPad Prizm. The B_{max} was estimated by dividing specific binding by fractional occupancy, where specific binding was total minus non-specific and fractional occupancy was ([radioligand]/(K_d + [radioligand]).

activation process (18). Treatment of sea urchin microsomes with low concentrations of NAADP seems to occlude the NAADP binding sites, or after binding of the radiolabel, the radioligand becomes 'trapped'. Therefore, in binding experiments using [^{32}P]NAADP, failure to add the radioactive substance simultaneously to the unlabelled ligand will result in abolishment of detectable total binding or of detectable displacement. The trapping of radioligand might be seen as a useful way of tagging the receptor thus aiding its purification, however, it is released upon solubilization of membranes with detergents such as CHAPS. All the pharmacological agents that inhibit NAADP-induced Ca^{2+} release do not appear to affect specific [^{32}P]NAADP binding to sea urchin microsomes (22).

References

1. Clapper, D., Walseth, T., Dargie, P., and Lee, H. C. (1987). *J. Biol. Chem.*, **262**, 9561.
2. Galione, A., Lee, H. C., and Busa, W. B. (1991). *Science*, **253**, 1143.
3. Lee, H. C. (1993). *J. Biol. Chem.*, **268**, 293.
4. Graeff, R. M., Podein, R. J., Aarhus, R., and Lee, H. C. (1995). *Biochem. Biophys. Res. Commun.*, **206**, 786.
5. Galione, A. and Summerhill, R. S. (1996). In *Ryanodine receptors* (ed. V. Sorrentino), p. 52. CRC Press, Boca Raton.
6. Aarhus, R., Gee, K., and Lee, H. C. (1995). *J. Biol. Chem.*, **270**, 7745.
7. Walseth, T. F. and Lee, H. C. (1993). *Biochim. Biophys. Acta*, **1178**, 235.
8. Bailey, V. C., Fortt, S. M., Summerhill, R. J., Galione, A., and Potter, B. V. (1996). *FEBS Lett.*, **379**, 227.
9. Sethi, J. K., Empson, R. M., Bailey, V. C., Potter, B. V., and Galione, A. (1997). *J. Biol. Chem.*, **272**, 16358.
10. Meszaros, L. G., Bak, J., and Chu, A. (1993). *Nature*, **364**, 76.
11. Gromada, J., Jorgensen, T. D., and Dissing, S. (1995). *FEBS Lett.*, **360**, 303.

12. Jones, K. T., Cruttwell, C., Parrington, J., and Swann, K. (1998). *FEBS Lett.*, **437**, 297.
13. Wilding, M., Russo, G. L., Galione, A., Marino, M., and Dale, B. (1998). *Am. J. Physiol.*, **275**, C1277.
14. Lee, H. C., Aarhus, R., Graeff, R., Gurnack, M. E., and Walseth, T. F. (1994). *Nature*, **370**, 307.
15. Missiaen, L. and Cheek, T. R. (1992). *Trends Cell Biol.*, **2**, 6.
16. Genazzani, A. A. and Galione, A. (1997). *Trends Pharmacol. Sci.*, **18**, 108.
17. Genazzani, A. A., Wilson, H. L., and Galione, A. (1999). *J. Mar. Biol. Assoc. UK*, **79**, 323.
18. Aarhus, R., Dickey, D. M., Graeff, R. M., Gee, K. R., Walseth, T. F., and Lee, H. C. (1996). *J. Biol. Chem.*, **271**, 8513.
19. Genazzani, A. A., Empson, R. M., and Galione, A. (1996). *J. Biol. Chem.*, **271**, 11599.
20. Dickey, D. M., Aarhus, R., Walseth, T. F., and Lee, H. C. (1998). *Cell Biochem. Biophys.*, **28**, 63.
21. Genazzani, A. A. and Galione, A. (1996). *Biochem. J.*, **315**, 721.
22. Genazzani, A. A., Mezna, M., Dickey, D. M., Michelangeli, F., Walseth, T. F., and Galione, A. (1997). *Br. J. Pharmacol.*, **121**, 1489.
23. Chini, E. N., Beers, K. W., and Dousa, T. P. (1995). *J. Biol. Chem.*, **270**, 3216.
24. Lee, H. C. and Aarhus, R. (1997). *J. Biol. Chem.*, **272**, 20378.
25. Rusinko, N. and Lee, H. C. (1989). *J. Biol. Chem.*, **264**, 11725.
26. Ashamu, G. A., Galione, A., and Potter, B. V. L. (1995). *Chem. Commun.*, 1359.
27. Lee, H. C. and Aarhus, R. (1993). *Biochim. Biophys. Acta*, **1164**, 68.
28. Howard, M., Grimaldi, J. C., Bazan, J. F., Lund, F. E., Santosargumedo, L., Parkhouse, R. M. E., *et al.* (1993). *Science*, **262**, 1056.
29. Sethi, J. K., Empson, R. M., and Galione, A. (1996). *Biochem. J.*, **319**, 613.
30. Graeff, R. M., Walseth, T. F., Hill, H. K., and Lee, H. C. (1996). *Biochemistry*, **35**, 379.
31. Walseth, T. F., Wong, L., Graeff, R. M., and Lee, H. C. (1997). In *Methods in enzymology* Vol. 280, p. 287. Ed. McCormick, D. B., Suttie, J. W., and Wagner, C. Academic Press, New York.
32. Galione, A., White, A., Willmott, N., Turner, M., Potter, B. V., and Watson, S. P. (1993). *Nature*, **365**, 456.
33. Higashida, H., Yokoyama, S., Hashii, M., Taketo, M., Higashida, M., Takayasu, T., *et al.* (1997). *J. Biol. Chem.*, **272**, 31272.
34. Graeff, R. M., Walseth, T. F., and Lee, H. C. (1997). In *Methods in enzymology* Vol. 280, p. 230. Ed. McCormick, D. B., Suttie, J. W., and Wagner, C. Academic Press, New York.
35. Reyes-Harde, M., Empson, R., Potter, B. V., Galione, A., and Stanton, P. K. (1999). *Proc. Natl. Acad. Sci. USA*, **96**, 4061.
36. Guse, A. H., da Silva, C. P., Berg, I., Skapenko, A. L., Weber, K., Heyer, P., *et al.* (1999). *Nature*, **398**, 70.
37. Munshi, C. and Lee, H. C. (1997). *Protein Expr. Purif.*, **11**, 104.
38. Vu, C. Q., Lu, P. J., Chen, C. S., and Jacobson, M. K. (1996). *J. Biol. Chem.*, **271**, 4747.
39. Cancela, J. M., Churchill, G. C., and Galione, A. (1999). *Nature*, **398**, 74.
40. Walseth, T. F., Yuen, P. S. T., and Moos, M. C. (1991). In *Methods in enzymology* Vol. 195, p. 29. Ed. McCormick, D. B., Suttie, J. W., and Wagner, C. Academic Press, New York.
41. Lee, H. C. (1991). *J. Biol. Chem.*, **266**, 2276.
42. Wilson, H. L. (1997). D.Phil Thesis, University of Oxford.

Chapter 11
Measuring calcium extrusion

Alexei V. Tepikin
The Physiological Laboratory, The University of Liverpool, Crown Street, Liverpool L69 3BX, UK.

1 Introduction

Calcium extrusion mechanisms of the plasma membrane—calcium ATPases, Na^+/Ca^{2+} exchangers, and calcium release with the content of secretory granules are essential elements in maintaining cellular calcium homeostasis. Three techniques were developed recently for measurements of calcium extrusion at the single cell level: the calcium clamp technique (1), the droplet technique (2–5), and the technique that employs calcium indicators bound to dextrans (calcium-sensitive jam technique) (6–8). The calcium clamp technique was based on intracellular ionophoretic injection of calcium. The measurement of the calcium extrusion rate was based on the assumption that to maintain a stable elevated concentration of calcium in the cell the rate of calcium injection (calculated from ionophoteric current) should be equal to the rate of calcium extrusion. The disadvantage of this technique was that we were able to apply it to large cells only (neurons of snail). This chapter will describe the two techniques of measurements of calcium extrusion that were developed later—the droplet technique (2–5) and the technique that employs calcium indicators bound to dextrans (calcium-sensitive jam technique) (6–8).

The main idea behind the droplet technique was to place a cell in a very small volume of extracellular solution (*Figures 1* and *2*). The volume of extracellular solution should be so small that calcium extruded from even a single isolated cell would be capable of producing measurable changes of extracellular calcium concentration. The technique was originally developed to measure calcium extrusion from the giant neurons of a snail (2). The droplet technique was later modified and adapted for measurements of calcium extrusion produced by small, isolated mammalian cells (3–5). The technique was used to measure calcium extrusion by calcium pumps of the plasma membrane (3, 4, 9) and for measurements of exocytotic calcium release (8). The droplet technique allows synchronous measurements (*Figure 3*) of calcium extrusion into the droplet solution and intracellular free calcium concentration ($[Ca^{2+}]_i$). The information derived from droplet experiments allows the calculation of the absolute extrusion rate, the changes of total intracellular calcium concentration, and the buffering capacity of the cell cytoplasm. The major advantage of the droplet technique experiments over multicellular methods of calcium extrusion measurements (e.g. Ca^{45} measurements or measurements of calcium extrusion using a calcium indicator and a suspension of cells in a cuvette in a fluorimeter) is that calcium extrusion occurring during complex and unsynchronized cellular reactions, like calcium oscillations, could be monitored and intracellular calcium signals could be correlated with calcium extrusion. The major disadvantage of the droplet technique is that it is relatively laborious and requires some technical skill in handling very small quantities of solutions.

The calcium-sensitive jam technique utilizes heavy dextrans to which calcium indicators are attached. These compound probes serve a dual purpose; they change fluorescence allowing detection of calcium ions and they also (because of high molecular weight) serve as buffers limiting diffusion of calcium. As a result of such dual buffer-indicator action, calcium gradients around the plasma membrane of the cell exist long enough to be detected by confocal microscopy (6–8). The calcium-sensitive jam experiments are technically easier than the droplet experiments. The total calcium flux could be derived from calcium jam experiments using complex calculations but the precision of total calcium flux measurements is inferior to the droplet technique. The important advantage of calcium jam experiments is that distribution of calcium extrusion sites on the plasma membrane can be resolved at a subcellular level (7, 8). The calcium-sensitive jam technique was recently used to measure calcium extrusion at the level of isolated organelles. The calcium indicators bound to heavy dextrans were used to measure calcium extrusion from the nuclear envelope into the nucleoplasm (10, see also Chapter 7 of this book); another application was the measurement of calcium extrusion from single secretory granules (11). Recently, simultaneous measurements of calcium extrusion from individual granules and intragranular calcium were made using confocal microscopy and dextran-bound indicators (12).

2 The droplet technique

2.1 Instruments and materials for the droplet technique

The schematic drawing of the droplet technique experiment is shown in *Figure 1*. A cell loaded with calcium indicator is placed into a droplet of solution containing another calcium probe. The optical properties of these indicators are different, therefore synchronous measurements of intracellular and extracellular calcium concentrations are possible. The most often used combination of indicators was fura-2 as intracellular indicator and fluo-3 as a probe for extracellular calcium measurements. Other combinations of intracellular end extracellular indicators are shown in *Table 1*. It is very important that the surface of the cover-glass on which the droplet is formed is siliconized, otherwise the droplet spreads, covers a large area, and can not be easily manipulated. Siliconization of cover-glasses was performed by immersing for 2–5 min into Sigmacote

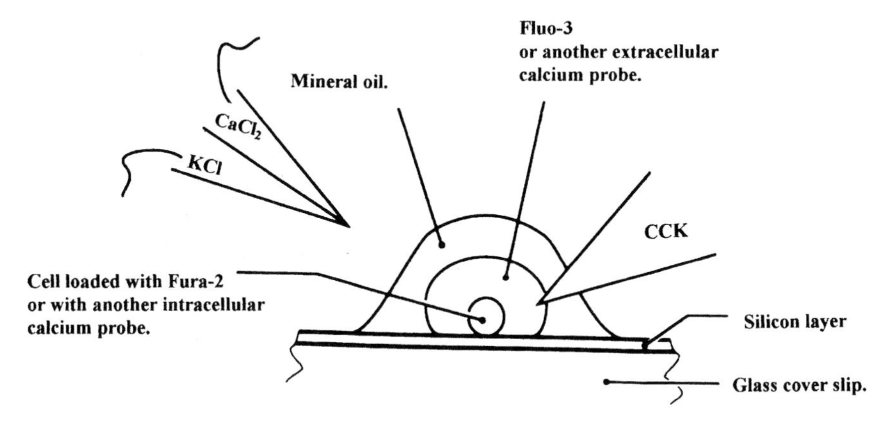

Figure 1 Schematic drawing of the set-up for a droplet experiment. Adapted from ref. 5, reproduced with permission from Springer–Verlag.

solution (Sigma). After removal from Sigmacote solution the cover-glass is rinsed with distilled water and dried. Siliconized cover-glasses are reusable for a few weeks. After every experiment the cover-glass should be washed with detergent and rinsed with distilled water, dried, and placed in a dust-free container.

Droplets were formed using a set of plastic pipettes. Two pipettes of different diameter are used for making the droplet. A coarse pipette with internal diameter of approx. 100 μm was used at initial stages of droplet formation. A smaller pipette with internal diameter 30–50 μm was used for the final adjustment of the droplet. These pipettes were produced from polyethylene tubing with internal diameter of 1–3 mm (Portex Limited, Kent, England). To make a pipette the central part of polyethylene tube (approx. 10 cm long) is melted by placing it over heating element (platinum wire connected to current source and heated to red-yellow colour); when the central part of the tube has melted the two ends of the tube are slowly pulled apart. This procedure is usually repeated twice for making pipettes of large diameter and three or four times for making pipettes suitable for the final stages of droplet formation.

Droplet volume is very small (usually 20–100 cellular volumes). A water droplet of this size would evaporate almost immediately if not covered with oil; paraffin oil from Fluka was used in most of our experiments. Mineral oil from Sigma (heavy white oil) is also suitable.

The extracellular droplet solution should be well pH buffered and calcium free. In our experiments the ionic composition of the extracellular droplet solution usually was: 110 mM NaCl, 4.7 mM KCl, 1.13 mM $MgCl_2$, 40 mM Hepes, 10 mM glucose pH 7.2.

2.2 Droplet formation

For combined intracellular/extracellular measurements cells should be loaded with a calcium-sensitive indicator (e.g. fura-2). For possible combinations of intracellular and extracellular indicators see *Table 1*.

Table 1 Combinations of calcium probes for droplet experiments

Intracellular indicator	Indicator in extracellular droplet solution
Fura-2	Fluo-3
Fura-2	Fluo-4
Fura-2	Rhod-2
Fura-2	Calcium Green-5N
Fura-2	Calcium ion-selective microelectrode
Fura-2	Antipyrylazo III
Fluo-3	Mg-fura-2

Protocol 1

Making droplet containing a single cell

Equipment and reagents

- Microscope
- Micromanipulator
- Siliconized cover-glass
- Pipettes of different diameter (see text)
- Mineral oil
- Calcium indicator (e.g. fluo-3)

Method

1. Place siliconized cover-glass on the microscope stage.
2. Place a large drop (1–3 μl) of cell suspension on the surface of the cover-glass (*Figure 2A*).
3. Cover immediately with mineral oil to prevent evaporation.
4. Using the microscope choose the best cell suitable for the experiment.
5. Use the coarse plastic pipette with tip diameter of approx. 100 μm to remove excess cells and to reduce the size of the droplet to approx. 300 μm in diameter. Use intermediate magnification of the microscope (total magnification of approx. × 40–60) for this procedure.
6. Wash the pipette with distilled water.
7. Use the same plastic pipette to infuse into the droplet the solution containing extracellular calcium indicator (e.g. 100 μM of fluo-3). The volume of droplet at this stage should increase more than 50 times.
8. Wait for a few minutes then once again use the pipette to reduce the size of droplet to approx. 300 μm.
9. Install the holder of the small tip diameter pipette into the micromanipulator. The holder used in our experiment was similar to a patch clamp pipette holder. Place the fine pipette (tip diameter 30–50 μm) into the holder.

Protocol 1 continued

10 Reduce the volume of the droplet by removal of solution and pressing the boundary of the droplet with the fine pipette. Suction applied to the pipette (similar to patch clamp experiments) allows the removal of solution from the droplet. The boundary of the droplet could be moved by pressing the droplet with the pipette. The combination of removal of solution and pipette movements allows the reduction of the diameter of the cell containing droplet to 60–140 μm (*Figure 2B*).

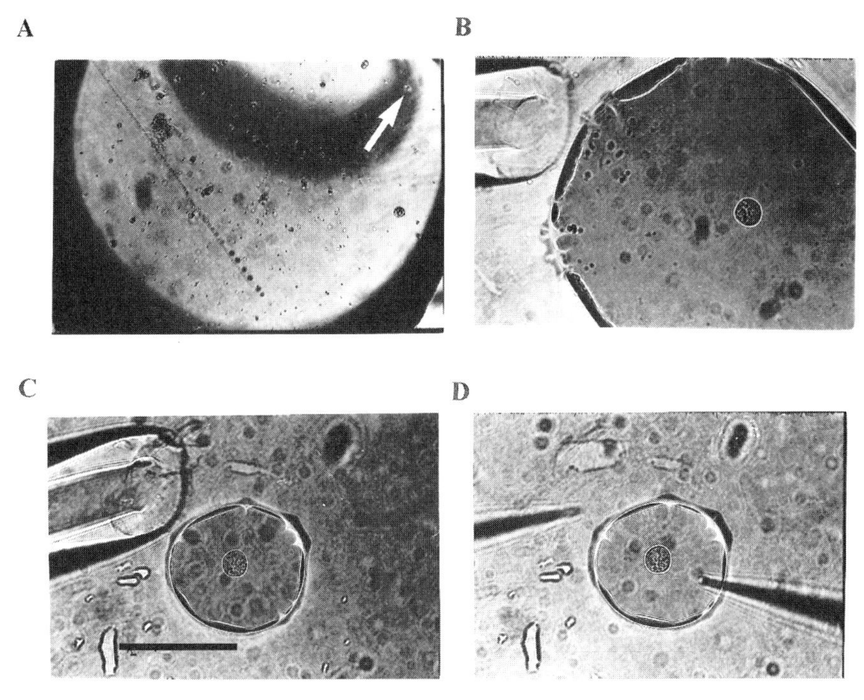

Figure 2 Droplet formation. (A) The initial drop contains many pancreatic acinar cells; the cell selected for the experiment is indicated by a white arrow. (B) Droplet at the beginning of final adjustment; the plastic pipette that was used for the final steps of droplet formation is shown to the right of this and the next picture. (C) Completely formed cell containing droplet. Scale bar for B–D represents 100 μm. (D) Cell containing droplet shown together with micropipette and microelectrode for iontophoretic injection. The micropipette in the right of the photograph contains agonist; the calcium containing two-barrelled microelectrode can be seen in the left of the photograph. Adapted from ref. 5, reproduced with permission of Springer–Verlag.

2.3 Injection of substances into the droplet

Pressure injection, iontophoretic injection or simply diffusion from a micropipette was used to stimulate cells in the droplet with different agonists. The washout of agonists in the droplet experiment is practically impossible. This is one of the disadvantages of the droplet technique. *Figures 1* and *2C* shows a cell containing droplet and glass micropipettes used to apply substances into the droplet. The volume delivered by the pressure injection procedure was calibrated by preliminary injections into the oil. In the droplet experiments the pressure was adjusted to deliver approx. 1% of the droplet volume. For pressure injection we used pipettes with internal tip diameter of approx. 1–2 μm. Pipettes were siliconized by immersing into Sigmacote solution. Pipettes were rinsed with distilled water and stored in a dust-free container before the experiment. Siliconization of large diameter pipettes is essential, otherwise, a substantial volume of droplet solution is removed by capillary forces during contact between the pipette and the droplet. In our experiments the hormone cholecystokinin (CCK) and thapsigargin (an inhibitor of calcium ATPases of internal stores) were delivered into cell containing droplets using pressure injection. Acetylcholine (ACh) was delivered into the droplets using both pressure injection and iontophoretic injection. In some experiments simple diffusion from micropipette was used do deliver CCK or ACh into the droplet. The micropipettes for the diffusion application were relatively sharp (the resistance of such micropipette when filled with 1 M KCl was approx. 40 MO). Pipettes for the diffusion application were not siliconized. The micropipettes were filled with solution of the same ionic composition as the droplet solution and contained an agonist in a concentration approximately one thousand times higher than expected to produce a cellular reaction. In some experiments the pipette was withdrawn immediately following the rise of intracellular and extracellular calcium concentrations. At the end of an experiment the fluorescent indicator in the droplet was saturated by iontophoretic injection of calcium (*Figure 3B*). This was necessary to calibrate extracellular fluorescence changes and to calculate extracellular calcium concentration. The calcium injection was performed from a two-barrelled microelectrode with one barrel filled with $CaCl_2$ (200 mM) and another filled with KCl (500 mM).

2.4 Measurements of fluorescence changes of intracellular and extracellular indicators and calibration of calcium concentrations in the droplet experiment

An example of the droplet technique experiment is shown in *Figure 3*. The cell was loaded with fura-2. The extracellular solution contained fluo-3. Trace 1 shows changes of fluorescence in the extracellular droplet solution (wavelength of excitation 490 nm, emission 530 nm). Trace 2 shows changes of fluorescence recorded at excitation 380 nm—the wavelength often used for fura-2 measurements. A decrease of fluorescence of fura-2 at 380 nm excitation corresponds to increase of intracellular Ca^{2+} concentration. These changes of fluorescence indicate that an intracellular calcium rise induces calcium extrusion into the extracellular droplet solution. There is a small contribution of fluo-3 fluorescence at 380 nm excitation (trace 3) which should be subtracted from the combined fluorescence (see *Protocol 2*). The corrected record of fluorescence of intracellular fura-2 (at 380 nm excitation) is shown as a trace 4. Fura-2 is non-fluorescent at excitation 490 nm. The fluo-3 is a true single-wavelength indicator. The propor-

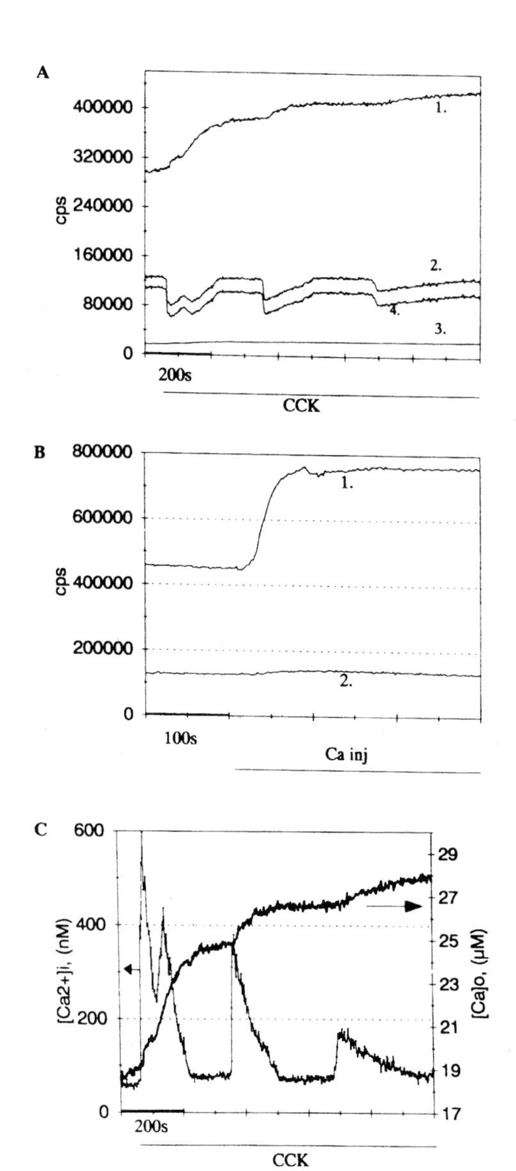

Figure 3 Recording of calcium changes in the cell, loaded with fura-2, and in extracellular droplet solution containing fluo-3. The cell was stimulated with cholecystokinin (CCK). (A) Changes of fluorescence of the cell/droplet system. 1 – Intensity of fluorescence recorded at 490 nm wavelength of excitation (fluorescence of extracellular fluo-3). 2 – Fluorescence recorded at 380 nm of excitation. 3 – Calculated intensity of fluorescence of fluo-3 at 380 nm of excitation. 4 – Calculated intensity of fluorescence of fura-2 loaded cell at 380 nm of excitation. (B) Saturation of extracellular droplet solution with calcium injected from a microelectrode (period of injection is shown by bar). 1 – Changes of fluorescence recorded at 490 nm excitation. 2 – Changes of fluorescence recorded at 380 nm excitation. (C) Calculated changes of intracellular and extracellular calcium concentrations. Reproduced from ref. 5, with permission of Springer–Verlag.

tional changes of fluorescence at 340 and at 380 nm of excitation are the same as at 490 nm excitation, in other words the ratio of fluorescence at different wavelengths of excitation does not change with changes of calcium concentration. This allows us to reconstitute the 380 nm and 340 nm changes of fluo-3 fluorescence from the 490 nm fluorescence record. *Figure 3B* shows the saturation of extracellular fluo-3 with calcium. This allows calculation of calcium concentration in the droplet solution. *Figure 3C* shows the calculated intracellular and extracellular calcium changes for this droplet experiment. Please note that for the droplet solution total calcium concentration is calculated whilst the intracellular recording shows free calcium concentration.

Protocol 2

Measuring intensity of fluo-3 fluorescence at different wavelengths of excitation

Equipment and reagents

- Fluorescence microscope
- Cover-glass
- Mineral oil
- Fluo-3
- $CaCl_2$

Method

1. Add fluo-3 to a small volume (e.g.100 μl) of your standard extracellular droplet solution. The final concentration of fluo-3 should be similar to what you are planning to use in the droplet experiments (e.g. 100 μM).
2. Add $CaCl_2$ to the fluo-3 containing solution to achieve the final total concentration of calcium of 1 mM (this will almost completely saturate fluo-3).
3. Place 1 μl drop of this solution on the cover-glass and cover immediately with mineral oil.
4. Place the cover-glass with oil covered droplet on the stage of your fluorescence microscope.
5. Measure the intensity of fluorescence at 340 nm, 380 nm, and 490 nm wavelengths of excitation: F_{fluo}^{340}, F_{fluo}^{380}, F_{fluo}^{490}.
6. Find the ratios $\alpha^{380/490} = F_{fluo}^{380}/F_{fluo}^{490}$ and $\alpha^{340/490} = F_{fluo}^{340}/F_{fluo}^{490}$.

The coefficients determined using *Protocol 2* are used to determine and subtract the contribution of fluo-3 fluorescence (see *Protocol 3*).

In droplet experiments we measured and calculated total extracellular calcium concentrations in the droplet solution. In this case we are actually interested in total and not free calcium concentration since the measurements of the total concentration allow us to calculate calcium fluxes that are produced by calcium extrusion mechanisms of single isolated cells.

Protocol 3

Calculations of intracellular calcium concentrations in droplet experiments

Equipment and reagents

- See *Protocol 2*

Method

1 Record the fluorescence of the cell-droplet system at excitation wavelengths 340 nm, 380 nm, and 490nm (correspondingly F_{total}^{340}, F_{total}^{380}, F_{total}^{490}). A common emission filter centred at 530 nm is suitable for all three types of measurements. Please note that because of the negligible contribution of fura-2 fluorescence at 490 nm excitation, F_{total}^{490} is equal to F_{fluo}^{490}. This protocol is written under the assumption that the cell is loaded with fura-2 and the extracellular solution contains fluo-3 (the protocol should be the same for recently developed single-wavelength indicators like Calcium Green-1 or fluo-4).

2 To find the fluorescence of intracellular fura-2 at the excitation wavelength 380 nm (F_{fura}^{380}) and at the excitation wavelength 340 nm (F_{fura}^{340}) one should subtract the contribution of fluo-3 from the 340 nm trace and from the 380 nm trace using the 490 nm fluorescence record and the previously determined (see *Protocol 2*) coefficients, $\alpha^{340/490}$ and $\alpha^{380/490}$:

$$F_{fura}^{380} = F_{total}^{380} - F_{fluo}^{380} = F_{total}^{380} - \alpha^{380/490} \times F_{fluo}^{490}$$

$$F_{fura}^{340} = F_{total}^{340} - F_{fluo}^{340} = F_{total}^{340} - \alpha^{340/490} \times F_{fluo}^{490}$$

3 Calculate the ratio $R = F_{fura}^{340}/F_{fura}^{380}$ for each of the time points. Use the standard formula for calculation of intracellular calcium concentration (13, 14, also see Chapter 2 of this book). Calibration of fura-2 (e.g. determining ratios and intensities of fluorescence that correspond to calcium-free and calcium-bound forms of indicator) should be performed in a separate experiment on cells placed in a perfused chamber.

Protocol 4

Measurements of extracellular calcium concentrations using the droplet technique

Equipment and reagents

- See *Protocol 2*
- Two-barrelled microelectrode

Method

1 Measure fluo-3 fluorescence at 490 nm excitation (F_{fluo}^{490}) in the extracellular droplet solution during your experiment.

Protocol 4 continued

2 Inject Ca^{2+} into the droplet solution at the end of experiment using a two-barrelled microelectrode. An injection current of 10–100 nA is sufficient to quickly saturate the indicator in the droplet solution with calcium.
3 Find the maximal value of fluorescence attained as a result of calcium injection. The maximal value of fluorescence determined in this way is termed $F_{fluo}{}^{490}{}_{max}$.
4 Use measured values of fluo-3 fluorescence ($F_{fluo}{}^{490}$) to find the total calcium concentration values at each time point of experiment:

$$[Ca]_o = K_{d,fluo} \times F_{fluo}{}^{490} / (F_{fluo}{}^{490}{}_{max} - F_{fluo}{}^{490}) + [fluo\text{-}3] \times F_{fluo}{}^{490} / F_{fluo}{}^{490}{}_{max}$$

where [fluo-3] is total concentration of the indicator in the extracellular solution and $K_{d,fluo}$ is the dissociation constant.

The method of calculation described in *Protocol 4* is based on the assumption that the fluorescence of the calcium-free form of the indicator is negligible. This assumption is true for many single-wavelength indicators. A more general equation for the calculation of total calcium concentration in the extracellular droplet solution is:

$$[Ca]_{dr} = K_d (F - F_{min}) / (F_{max} - F) + [indicator] (F - F_{min}) / (F_{max} - F_{min})$$

where F_{min} is the fluorescence of the calcium-free form of the indicator, F_{max} is the fluorescence of the calcium-bound form of the indicator (5).

2.5 Measuring droplet volume and volume of cells: calculation of calcium flux, rate of intracellular calcium concentration changes, and calcium binding capacity of the cytoplasm of the cell

In our experiments we used a calibrated ocular to measure cellular diameter. The volume of the cell (or cells) in the droplet was estimated in assumption of their spherical shape.

Protocol 5

Measuring the volume of the droplet

Equipment and reagents

- Small diameter plastic pipette (see *Protocol 1*)
- Cover-glass
- Calibrated ocular
- Mineral oil

Method

1 Apply suction to the small diameter plastic pipette (with tip diameter 30–50 μm) and use it to take the droplet into the pipette.

Protocol 5 continued

2 Lift the pipette (approx. 100 μm) above the surface of the cover-glass.

3 Expel the solution from the pipette. A spherical droplet will be formed inside the oil. This droplet will slowly descend towards the surface of the cover-glass.

4 Use a calibrated ocular to measure the diameter of the spherical droplet.

5 Calculate the volume of the droplet.

Changes of the amount of calcium in the droplet can be calculated as:

$$\Delta\ Ca_{dr} = \Delta\ [Ca]_{dr} \times V_{dr}$$

This, of course, is equal to the amount of calcium lost by the cell.
Changes of intracellular calcium concentration can be calculated using the following formula:

$$\Delta\ [Ca]_{cell} = \Delta\ [Ca]_{dr} \times V_{dr}\ /\ V_{cell}$$

The rate of intracellular calcium changes can be obtained by differentiating the cellular calcium curve. This can be done using numerical differentiation in Excel (Microsoft). Please note that changes in both total ($\Delta\ [Ca]_{cell}$) and free ($\Delta\ [Ca^{2+}]_{cell}$) intracellular calcium can be measured in droplet experiments. These measurements are sufficient to calculate the calcium binding capacity of the cytoplasm of the cell (k_c):

$$k_c = \Delta\ [Ca]_{cell}\ /\ \Delta\ [Ca^{2+}]_{cell}$$

We used the declining part of calcium transients induced by supramaximal doses of the agonist CCK to evaluate the calcium binding capacity of pancreatic acinar cells (15). The important requirements for this application of the droplet technique is that the calcium extrusion should be the main mechanism responsible for the decline in free calcium concentration in the cell ($[Ca^{2+}]_{cell}$). An important correction that should be made for such measurements is correction for contribution of the intracellular indicator. The contribution of the calcium indicator to the calcium binding capacity of the cell can be calculated using formula derived by Neher and Augustine (16). The concentration of calcium indicator (necessary for this calculation) can be estimated using confocal microscopy of the cells loaded with indicator and small droplets containing the same indicator at known concentrations. The measurements should be performed using the wavelength corresponding to the isobestic point of an indicator (for single-wavelength indicators the comparison should be carried out for calcium saturated conditions).

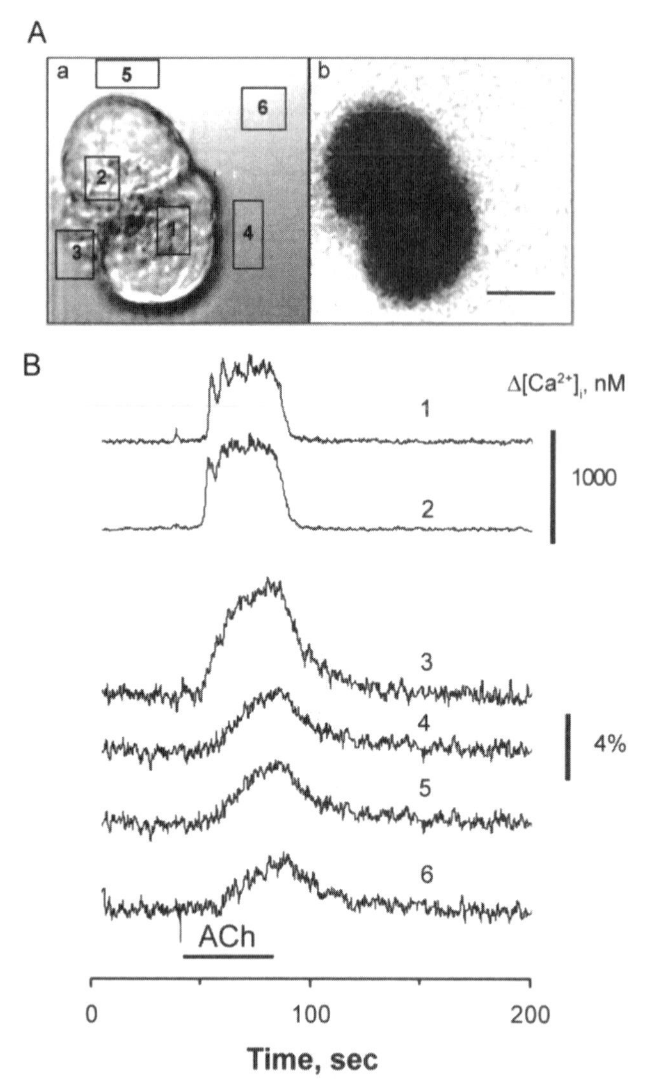

Figure 4 ACh-induced calcium extrusion from a pancreatic acinar cell cluster. (Aa) Transmitted light image of the cluster. The average fluorescence intensities in boxes 1–6 are displayed in (B) as a function of time. (Ab) Fluorescence intensity image at the emission wavelength of 530 nm using an optical slice going through the cluster. Fluorescence of fura red loaded into the cells of the cluster is not seen because the measurements of the emission light was at 530 nm (far from the emission maximum of fura red which is 650 nm). The scale bar represents 15 μm. (B) Simultaneous measurements of Ca^{2+} extrusion and changes in $[Ca^{2+}]_i$ in response to an iontophoretic ACh application to the cluster shown in (A). The upper traces (1 and 2) show free calcium concentration inside the cells of the cluster. The lower traces represent the fluorescence intensity of Calcium Green-1dextran in the extracellular solution. The number near a trace corresponds to the number of the box from which the trace was obtained. The duration of ACh application is indicated by the horizontal bar. The lower vertical bar in the right of the figure shows the relative changes of the fluorescence intensity of Calcium Green-1 dextran ($\Delta F/Fo$). Off-set in the vertical direction (here and in *Figure 5*) between traces was introduced to improve the clarity via trace separation. Reproduced from ref. 7, with permission from Pearson Professional Ltd.

3 Calcium-sensitive jam technique—using calcium indicators bound to dextrans to measure calcium extrusion

3.1 Instruments and materials for calcium-sensitive jam technique

There are two important requirements for the calcium-sensitive jam technique measurements. First—the dextran to which a calcium indicator is bound should be of high molecular weight (e.g. a dextran with molecular weight of 500 000 gives substantially better results than a dextran with molecular weight of 70 000). Secondly—confocal microscopy should be used to measure the fluorescence changes of dextran-bound indicators. The changes of fluorescence of a dextran-bound indicator is usually relatively small (a few per cent in our experiments), the out-of-focus fluorescence of the extracellular solution reduces the signal-to-noise ratio and makes measurements practically impossible. Calcium Green-1 dextran with molecular weight 500 000 was used in our experiments for measurements of extracellular calcium concentration, and fura red was used for measuring the intracellular calcium concentration. The difference in emission spectra of the two indicators allow simultaneous recordings of intracellular and extracellular calcium concentrations (see *Figures 4* and 5). The measurements of intracellular and extracellular calcium concentrations could be combined with uncaging of caged calcium (*Figure* 5). Estimation of the calcium diffusion coefficient in the presence of different concentrations of dextran could be made using formulae described by Belan and co-authors (6). We found empirically that the best results for extracellular calcium measurements for experiments with single cells (pancreatic acinar cells) or small cellular clusters (three to ten cells) are obtained with a concentration of Calcium Green-1 (bound to dextran) in the range 30–100 μM. It is important to note that the concentration of the dextran and concentration of indicator bound to dextran is different. Each molecule of dextran carries a number of molecules of indicator and this number depends on the particular batch of the probe. We found that the level of calcium contamination of the probe can vary considerably. In some cases the contamination had to be reduced. Calcium Sponge S (Molecular Probes) was used in our experiments to reduce calcium contamination of the probe. In our experiments the extracellular free calcium concentration was usually in the range 0.2–0.4 μM.

Cells in the chamber containing Calcium Green-1 dextran could be stimulated by simply adding a droplet of concentrated agonist solution into the chamber. However, in the majority of our experiments we used iontophoretic injection (of ACh) to stimulate the cells. The advantage of iontophoretic injection is that other cells in the chamber are not affected when a particular cell is stimulated. Calcium Green-1 dextran is relatively expensive and it is important to have the possibility to carry out a number of experiments using the same batch of cells and the same extracellular solution containing the indicator. To apply ACh we used a sharp microelectrode that was filled with ACh (1–10 mM). The injection

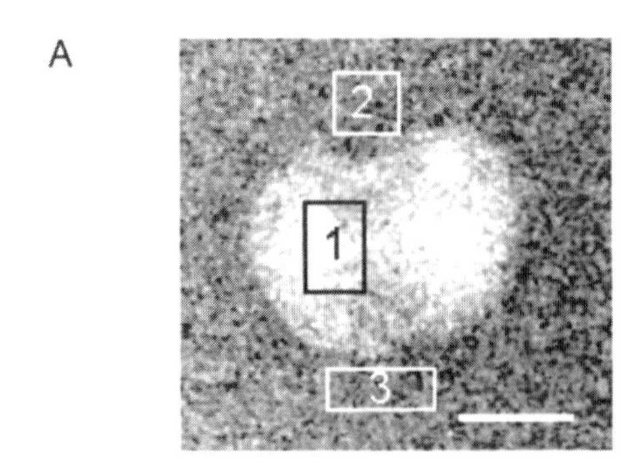

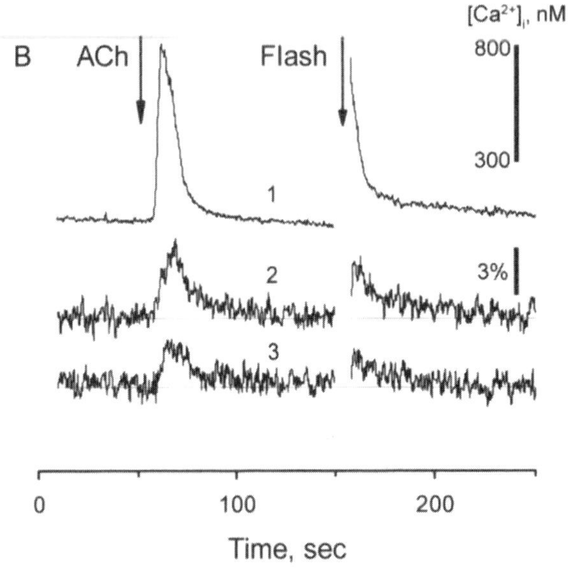

Figure 5 Calcium efflux induced by consecutive ACh application and photolysis of NP-EGTA. (A) Fluorescence image of the cell cluster. The cells were loaded with fura red. The extracellular solution contained Calcium Green-1 dextran. The scale bar represents 15 μm. The fluorescence intensities measured in boxes 1–3 are displayed in (B) as a function of time. (B) The fluorescence intensities measured in boxes 1–3, are shown in (A). The moment of ACh application and the time point of the UV light flash are shown by arrows. The upper trace represents the intracellular free calcium concentration changes whereas the lower ones represent the calcium concentration changes in the extracellular solution. The scale of fluorescence intensity changes of Calcium Green-1 dextran is indicated by the vertical bar next to the traces. Reproduced from ref. 7, with permission from Pearson Professional Ltd.

current was in the range 10–100 nA (most frequently 10–20 nA) the retaining current was 5–10 nA. The tip of the microelectrode was positioned 10–50 μm from the cell membrane.

Another way of inducing a calcium reaction in individual cells of the population is to uncage caged calcium (*Figure 2*). In our experiments we used NP-EGTA (Molecular Probes). Cells were loaded with NP-EGTA by incubating (for 30 min at room temperature) in a solution containing 5 μM of NP-EGTA, AM. UV laser light (364 nm) was used for uncaging in our experiments.

Protocol 6

Measuring calcium extrusion with Calcium Green-1 dextran and synchronous measurements of intracellular calcium changes with fura red

Equipment and reagents

- Confocal microscope
- Cover-glass
- Calcium indicator, e.g. fura red
- Calcium-free extracellular solution
- Calcium Green-1 bound to dextran (molecular weight 500 000)
- 1 M $CaCl_2$ solution
- Calcium ionophore, e.g. ionomycin

Method

1. Load the cells with the intracellular calcium indicator fura red (by incubation with fura red/AM in our experiments).
2. Wash cells twice in nominally calcium-free extracellular solution.
3. Place a drop (approx. 100 μl) of suspension of loaded cells in calcium-free solution onto a small (approx. 200 μl capacity) cover-glass bottomed chamber. In some experiments a small concentration of EGTA (20–50 μM) was added to the calcium-free solution to buffer calcium contamination.
4. Place the chamber on the table of the confocal microscope (an inverted microscope was used in our experiments). The confocal section should go through the middle of the cell—this will minimize the cross-talk between the two indicators.
5. Wait a few minutes for cells to adhere to the cover-glass.
6. Slowly remove the bulk of extracellular solution using a 100 μl pipette.
7. Add 200 μl of nominally calcium-free solution with 30–100 μM of Calcium Green-1 bound to dextran added.
8. Use a 488 nm laser line to excite both intracellular and extracellular indicators.
9. Use a 520–530 nm centred filter to record extracellular calcium changes, use a long pass filter (e.g. > 620 nm) to record intracellular calcium changes.
10. At the end of the experiment saturate the extracellular indicator by adding a small volume of concentrated (1 M) $CaCl_2$ solution to the extracellular solution (the final concentration of $CaCl_2$ should be 1–5 mM).
11. Record the maximal fluorescence of the extracellular solution that develops after addition of $CaCl_2$. Use this value to calculate calcium changes in extracellular solution that occurred during the experiment.
12. After the extracellular solution is saturated add ionomycin (a calcium ionophore) to the extracellular solution (final concentration 20 μM) to saturate and calibrate the intracellular calcium indicator.

3.2 Calcium-sensitive gel

To make indicators bound to dextrans even less mobile they can be incorporated into agarose gel (P. Thorn and A. Tepikin, unpublished data). The gel was made from a solution containing 1% of low gelling temperature agarose (Agarose type IX from Sigma was used in our experiments). 100 μM of indicator (Calcium Green-1 bound to dextran) was added to this solution before gel formation. The gel forms around the cells and the cells survive in such a gel for many tens of minutes. One problem with these types of experiments is that agarose is heavily contaminated with calcium. To reduce calcium contamination non-polymerized agarose was washed (three or four times) by centrifugation with nominally calcium-free EGTA containing solution. The last wash (before melting) was with nominally calcium-free solution without EGTA. The agarose was melted by heating to approximately 60 °C, and then allowed to cool. The indicator–dextran was added to the melted agarose when the temperature had dropped to 25–30 °C. The pancreatic acinar cells were suspended in a solution containing melted agarose and a dextran-bound calcium probe. 200 μl of this cell-containing solution was then placed on the surface of the cover-glass and further cooled to approximately 15 °C. There is a considerable hysteresis of liquid–gel transition of agarose solutions; agarose gel with this content of agarose will not re-melt unless it is heated to more than 50 °C. The experiments were performed at room temperature. Pilot experiments have shown that this technique for measuring calcium extrusion is very sensitive and allows the use of indicators bound to smaller dextrans (with molecular weight of 70 000). The disadvantage of this technique is that it is more laborious than simple indicator–dextran based measurements.

References

1. Belan, P. V., Kostyuk, P. G., Snitsarev, V. A., and Tepikin, A. V. (1993). *J. Physiol.*, **462**, 47.
2. Tepikin, A. V., Kostyuk, P. G., Snitsarev, V. A., and Belan P. V. (1991). *J. Membr. Biol.*, **123**, 43.
3. Tepikin, A. V., Voronina, S. G., Gallacher, D. V., and Petersen, O. H. (1992). *J. Biol. Chem.*, **267**, 3569.
4. Tepikin, A. V., Voronina, S. G., Gallacher, D. V., and Petersen, O. H. (1992). *J. Biol. Chem.*, **267**, 14073.
5. Tepikin, A. V., Llopis, J., Snitsarev, V. A., Gallacher, D. V., and Petersen, O. H. (1994). *Pflugers Arch.*, **428**, 664.
6. Belan, P. V., Gerasimanko, O. V., Berry, D., Saftenku, E., Petersen, O. H., and Tepikin, A. V. (1996). *Pflugers Arch.*, **433**, 200.
7. Belan, P., Gerasimenko, O., Petersen, O. H., and Tepikin, A. V. (1997). *Cell Calcium*, **22**, 5.
8. Belan, P., Gerasimenko, O., Petersen, O. H., and Tepikin, A. V. (1998). *J. Biol. Chem.*, **273**, 4106.
9. Camello, P., Gardner, J., Petersen, O. H., and Tepikin, A. V. (1996). *J. Physiol.*, **490**, 585.
10. Gerasimenko, O. V., Gerasimenko, J. V., Tepikin, A. V., and Petersen, O. H. (1995). *Cell*, **80**, 439.
11. Gerasimenko, O. V., Gerasimenko, J. V., Belan, P. V., and Petersen, O. H. (1996). *Cell*, **84**, 473.

12. Nguyen, T., Chin, W. C., and Verdugo, P. (1998). *Nature*, **395** (6705), 908.
13. Grynkiewicz, G., Poenie, M., and Tsien, R. Y. (1985). *J. Biol. Chem.*, **260** (6), 3440.
14. Thomas, A. P. and Delaville, F. (1991). In *Cellular calcium: a practical approach* (ed. J. G. McCormack and P. H. Cobbold), p. 1. IRL Press at Oxford University Press, Oxford.
15. Mogami, H., Gardner, J., Gerasimenko, O. V., Camello, Petersen, O. H., and Tepikin, A. V. (1999). *J. Physiol.*, **518** (2), 463.
16. Neher, E. and Augustine, G. J. (1992). *J. Physiol.*, **450**, 273.

Appendix 1
Fluorescent calcium indicators and caged calcium probes discussed in this volume

The data is based on information from Molecular Probes website (*www.probes.com*), TefLabs website (*www.teflabs.com*), and information from the contributors of this book. The wavelengths indicated for excitation and emission of particular indicators are usable but should be optimized for each individual type of measurements.

1 Fluorescent indicators

Indicators (in alphabetical order)	K_d for Ca^{2+}	Excitation (nm) Emission (nm)	Source	Comments	Chapter Protocol
Calcium Crimson	185 nM	590 ex 615 em	Molecular Probes	Single-wavelength indicator, fluorescence increases upon calcium binding. Long wavelengths of excitation and emission.	Chapter 3
Calcium Green-1	190 nM	488 ex 530 em	Molecular Probes	Single-wavelength indicator, fluorescence increases upon calcium binding, used to form indicator–dextran complexes (see below), suitable for confocal microscopy.	Chapter 11
Calcium Green-1 dextran (MW 70 000)	240 nM	488 ex 530 em	Molecular Probes	Indicator–dextran complex, used for measurements of calcium in nucleoplasm of isolated nuclei.	Chapter 7. Protocol 3 in Chapter 7.
Calcium Green-1 dextran (MW 500 000)	310 nM	488 ex 530 em	Molecular Probes	Indicator–dextran complex, used to measure the distribution of calcium extrusion at cellular and subcellular level.	Chapter 11
Calcium Green-2	550 nM	490 ex 540 em	Molecular Probes	Intermediate affinity calcium indicator. Fluorescence increases upon calcium binding.	Chapter 3
Calcium Green-5N	14 μM	488 ex 530 em	Molecular Probes	Low affinity calcium indicator used for measurements of calcium extrusion, fluorescence increases upon calcium binding.	Chapter 11

Indicators (in alphabetical order)	K_d for Ca^{2+}	Excitation (nm) Emission (nm)	Source	Comments	Chapter Protocol
Calcium Orange	185 nM	540 ex 580 em	Molecular Probes	Single-wavelength indicator, fluorescence increases upon calcium binding.	Chapter 3
Fluo-3	390 nM	488 ex 530 em	Molecular Probes and range of other companies including Sigma and Calbiochem	Single-wavelength indicator with strong increase of emission upon calcium binding, used for intracellular and extracellular calcium measurements. Popular indicator for confocal measurements.	Chapters 2,5, 9, 10, 11. Protocols 2 and 4 in Chapter 11.
Fluo-3 FF	41 μM	515 ex 530 em	TefLabs	Low affinity, used for ER calcium measurements, fluorescence increases upon calcium binding.	Chapter 6
Fluo-4	345 nM	488 ex 520 em	Molecular Probes	Similar to fluo-3 but brighter and with larger increase of fluorescence upon calcium binding. Becoming a very popular indicator for confocal measurements.	Chapters 2, 11
Fura-2	135 nM (22°C) 224 nM (37°C)	340/380 ex 510 em	Molecular Probes and range of other companies including Sigma and Calbiochem	Probably the most frequently used indicator for cytosolic calcium measurements, high affinity, ratiometric (on excitation).	Chapters 2, 3, 5, 7, 8, 9, 11. Protocol 5 in Chapter 2. Protocol 3 in Chapter 3. Protocol 5 in Chapter 9. Protocol 3 in Chapter 11.
Fura-2 FF	35 μM	340/380 ex 510 em	TefLabs	Low affinity, used for ER calcium measurements, ratiometric (on excitation).	Chapter 6
Fura-PE3	250 nM	340/380 ex 510 em	TefLabs	Leakage resistant derivative of fura-2 (see above).	Chapters 2, 8. Protocol 5 in Chapter 8.
Fura red	140 nM	488 ex 655 em	Molecular Probes	Fluorescence decreases upon calcium binding. Used in confocal experiments. Can be combined with fluo-3 for ratiometric confocal measurements.	Chapters 2, 7
Indo-1	230 nM	360 ex 405/495 em	Molecular Probes and range of other companies including Sigma and Calbiochem	Ratiometric (on emission) calcium probe. Extensively used in calcium measurements in muscle cells. Suitable for ratiometric confocal calcium measurements.	Chapters 2, 9

Indicators (in alphabetical order)	K_d for Ca^{2+}	Excitation (nm) Emission (nm)	Source	Comments	Chapter Protocol
Mag-fura-2 (furaptra)	25 μM*	345/375 ex 510 em	Molecular Probes	Low affinity, used for ER calcium measurements, ratiometric (on excitation).	Chapters 3, 6
Mag-fura-5	28 μM	340/380 ex 510 em	Molecular Probes	Low affinity, used for ER calcium measurements, ratiometric (on excitation).	Chapter 6
Mag-fura red	17 μM	488 ex 610 em	Molecular Probes	Low affinity, used for ER calcium measurements, fluorescence decreases upon calcium binding.	Chapter 6
Mag-Indo-1	32 μM	351 ex 405/485 em	Molecular Probes	Low affinity, used for ER calcium measurements, ratiometric (on emission).	Chapter 6
Oregon Green BAPTA-1N	170 nM	488 ex 520 em	Molecular Probes	Excited efficiently by 488 nm argon laser line. Single-wavelength indicator, fluorescence increases upon calcium binding.	Chapter 3
Oregon Green BAPTA-5N	20 μM	492 ex (488 ex) 521 em	Molecular Probes	Low affinity, used for ER calcium measurements, fluorescence increases upon calcium binding; excited efficiently by 488 nm argon laser line.	Chapter 6
Rhod-2	570 nM	555 ex 600 em	Molecular Probes	Positively charged in AM form. Accumulate in mitochondria, used for mitochondrial calcium measurements. Fluorescence increases upon calcium binding.	Chapters 3, 5, 11. Protocol 1 in Chapter 5.

2 Caged calcium probes

Cage probe	K_d for calcium in caged form	K_d for calcium in uncaged form	Source	Comments	Chapter
NP-EGTA	80 nM	1 mM	Molecular Probes	Low affinity for magnesium. Decreases affinity for calcium upon UV illumination. Photolysis is faster than that of DM-nitrophen (see below).	Chapter 11
DM-nitrophen	5 nM	3 mM	Calbiochem Molecular Probes (as DMNP-EDTA)	High affinity for magnesium in caged form. Decreases affinity to calcium (and magnesium) upon UV illumination. Used for two-photon uncaging of calcium.	Chapter 2

*Larger value (53 μM) is given in Chapter 6.

List of suppliers

Abbot Laboratories, Abbot Park, IL, USA.
URL: http://www.abbot.com

Anderman and Co. Ltd., 145 London Road, Kingston-upon-Thames, Surrey KT2 6NH, UK.
Tel: 0181 541 0035 Fax: 0181 541 0623

Applied Scientific (supplier of Irvine Scientific Products), 154 West Harris Avenue, South San Francisco, CA 94080, USA.
URL: http://www.irvinesci.com

Aurora Bioscience, 11010 Torreyana Road, San Diego, CA, USA.
URL: http://www.aurorabio.com

Axon Instruments, Inc., 1101 Chess Drive, Foster City, CA 94404, USA.
URL: http://www.axon.com

Baxter Healthcare Corporation, I.V. Systems Division, Deerfield, IL 60015, USA.
URL: http://www.baxter.com

BDH Inc., 350 Evans Avenue, Toronto, Ontario M8Z 1K5, Canada.
URL: http://www.bdhinc.com

Beckman Coulter (UK) Ltd., Oakley Court, Kingsmead Business Park, London Road, High Wycombe, Buckinghamshire HP11 1JU, UK.
Tel: 01494 441181
Fax: 01494 447558
URL: http://www.beckman.com
Beckman Coulter Inc., 4300 N Harbor Boulevard, PO Box 3100, Fullerton, CA 92834-3100, USA.
Tel: 001 714 871 4848
Fax: 001 714 773 8283
URL: http://www.beckman.com

Becton Dickinson and Co., 21 Between Towns Road, Cowley, Oxford OX4 3LY, UK.
Tel: 01865 748844
Fax: 01865 781627
URL: http://www.bd.com
Becton Dickinson and Co., 1 Becton Drive, Franklin Lakes, NJ 07417-1883, USA.
Tel: 001 201 847 6800
URL: http://www.bd.com

Bio 101 Inc., c/o Anachem Ltd., Anachem House, 20 Charles Street, Luton, Bedfordshire LU2 0EB, UK.
Tel: 01582 456666
Fax: 01582 391768
URL: http://www.anachem.co.uk
Bio 101 Inc., PO Box 2284, La Jolla, CA 92038-2284, USA.
Tel: 001 760 598 7299
Fax: 001 760 598 0116
URL: http://www.bio101.com

Bio-Rad Laboratories Ltd., Bio-Rad House, Maylands Avenue, Hemel Hempstead, Hertfordshire HP2 7TD, UK.
Tel: 0181 328 2000 Fax: 0181 328 2550
URL: http://www.bio-rad.com

Bio-Rad Laboratories Ltd., Division Headquarters, 1000 Alfred Noble Drive, Hercules, CA 94547, USA.
Tel: 001 510 724 7000
Fax: 001 510 741 5817
URL: http://www.bio-rad.com

Cairn Research Ltd., Unit 3G, Brents Shipyard Industrial Estate, Faversham, Kent ME13 7DZ, UK.
URL: http://www.cairnweb.com

Calbiochem–Novabiochem Corporation, 10394 Pacific Center Court, San Diego, CA 92121, USA.
URL: http://www.calbiochem-novabiochem.com

Carl Zeiss Microscopy, D-07740 Jena, Germany. URL: http://www.zeiss.de

Chroma Technology Corp., 72 Cotton Mill Hill, Unit A9, Brattleboro, VT 05301, USA.
URL: http://www.chroma.com

Cole-Parmer Instrument Co. Ltd., Unit 3, River Brent Business Park, Trumpers Way, Hanwell, London W7 2QA, UK.
Cole-Parmer Instrument Co., 625 East Bunker Court, Vernon Hills, IL 60061, USA.
URL: http://www.cole-parmer.com

CP Instrument Co. Ltd., PO Box 22, Bishop Stortford, Hertfordshire CM23 3DX, UK.
Tel: 01279 757711 Fax: 01279 755785
URL: http://www.cpinstrument.co.uk

Dupont (UK) Ltd., Industrial Products Division, Wedgwood Way, Stevenage, Hertfordshire SG1 4QN, UK.
Tel: 01438 734000 Fax: 01438 734382
URL: http://www.dupont.com
Dupont Co. (Biotechnology Systems Division), PO Box 80024, Wilmington, DE 19880-002, USA.
Tel: 001 302 774 1000
Fax: 001 302 774 7321
URL: http://www.dupont.com

Eastman Chemical Co., 100 North Eastman Road, PO Box 511, Kingsport, TN 37662-5075, USA.
Tel: 001 423 229 2000
URL: http://www.eastman.com

Fisher Scientific UK Ltd., Bishop Meadow Road, Loughborough, Leicestershire LE11 5RG, UK.
Tel: 01509 231166
Fax: 01509 231893
URL: http://www.fisher.co.uk
Fisher Scientific, Fisher Research, 2761 Walnut Avenue, Tustin, CA 92780, USA.
Tel: 001 714 669 4600
Fax: 001 714 669 1613
URL: http://www.fishersci.com
Fisher Scientific Co., 711 Forbest Avenue, Pittsburg, PA 15219-4785, USA.

Fluka Chemicals Ltd., The Old Brickyard, New Road, Gillingham, Dorset SP8 4JL, UK.
Fluka, PO Box 2060, Milwaukee, WI 53201, USA.
Tel: 001 414 273 5013
Fax: 001 414 2734979
URL: http://www.sigma-aldrich.com
Fluka Chemical Co. Ltd., PO Box 260, CH-9471, Buchs, Switzerland.
Tel: 0041 81 745 2828
Fax: 0041 81 756 5449
URL: http://www.sigma-aldrich.com

Gilson, Inc., 3000 W. Beltline Hwy., PO Box 620027, Middleton, WI 53562, USA.
Gilson, Inc., 72 rue Gambetta, BP 45, 95400 Villiers-le-Bel, France.
URL: http://www.gilson.com

Hamamatsu Corp. USA, 360 Foothill Road, Bridgewater, NJ 08807, USA.
URL: http://www.hamamatsu.com

Hitachi Instruments, 1832 Centre Point Drive, Suite 102, Naperville, IL 60563, USA.
URL: http://www.hitachi.com

Hybaid Ltd., Action Court, Ashford Road, Ashford, Middlesex TW15 1XB, UK.
Tel: 01784 425000 Fax: 01784 248085
URL: http://www.hybaid.com
Hybaid US, 8 East Forge Parkway, Franklin, MA 02038, USA.
Tel: 001 508 541 6918
Fax: 001 508 541 3041
URL: http://www.hybaid.com

HyClone Laboratories, 1725 South HyClone Road, Logan, UT 84321, USA.
Tel: 001 435 753 4584
Fax: 001 435 753 4589
URL: http://www.hyclone.com

Invitrogen Corp., 1600 Faraday Avenue, Carlsbad, CA 92008, USA.
Tel: 001 760 603 7200
Fax: 001 760 603 7201
URL: http://www.invitrogen.com
Invitrogen BV, PO Box 2312, 9704 CH Groningen, The Netherlands.
Tel: 00800 5345 5345
Fax: 00800 7890 7890
URL: http://www.invitrogen.com

Jobin Yvon / Spex, 16–18, rue du Canal, 91165 Longjumeau cedex, France.
URL: http://www.jobinyvon.com

Kinetic Imaging Ltd., South Harrington Building, Sefton Street, Liverpool L3 4BQ, UK.
URL: http://www.kineticimaging.com

Leica Microsystems Heidelberg GmbH, Im Neuenheimer Feld 518, D-69120, Germany.
URL: http://www.leica-microsystems.com

Life Technologies Ltd., PO Box 35, Free Fountain Drive, Incsinnan Business Park, Paisley PA4 9RF, UK.
Tel: 0800 269210 Fax: 0800 838380
URL: http://www.lifetech.com
Life Technologies Inc., 9800 Medical Center Drive, Rockville, MD 20850, USA.
Tel: 001 301 610 8000
URL: http://www.lifetech.com

Merck Sharp & Dohme, Research Laboratories, Neuroscience Research Centre, Terlings Park, Harlow, Essex CM20 2QR, UK.
URL: http://www.msd-nrc.co.uk
MSD Sharp and Dohme GmbH, Lindenplatz 1, D-85540, Haar, Germany.
URL: http://www.msd-deutschland.com

Millipore (UK) Ltd., The Boulevard, Blackmoor Lane, Watford, Hertfordshire WD1 8YW, UK.
Tel: 01923 816375
Fax: 01923 818297
URL: http://www.millipore.com/local/UK.htm
Millipore Corp., 80 Ashby Road, Bedford, MA 01730, USA.
Tel: 001 800 645 5476
Fax: 001 800 645 5439
URL: http://www.millipore.com

Molecular Probes, 4849 Pitchford Avenue, Eugene, OR, USA.
URL: http://www.probes.com

New England Biolabs, 32 Tozer Road, Beverley, MA 01915-5510, USA.
Tel: 001 978 927 5054
URL: http://www.neb.com

Nikon Inc., 1300 Walt Whitman Road, Melville, NY 11747-3064, USA.
Tel: 001 516 547 4200
Fax: 001 516 547 0299
URL: http://www.nikonusa.com
Nikon Corp., Fuji Building, 2-3, 3-chome, Marunouchi, Chiyoda-ku, Tokyo 100, Japan.
Tel: 00813 3214 5311
Fax: 00813 3201 5856
URL: http://www.nikon.co.jp/main/index_e.htm

Nycomed Amersham plc, Amersham Place, Little Chalfont, Buckinghamshire HP7 9NA, UK.
Tel: 01494 544000 Fax: 01494 542266
URL: http://www.amersham.co.uk

Nycomed Amersham, 101 Carnegie Center, Princeton, NJ 08540, USA.
Tel: 001 609 514 6000
URL: http://www.amersham.co.uk

Omega Engineering, PO Box 4047, Stamford, CT 06907-0047, USA.
URL: http://www.omega.com

Omega Optical, Inc., 3 Grove Street, PO Box 573, Brattleboro, VT 05302, USA.
URL: http://www.omegafilters.com

Oriel Instruments, 150 Long Beach Blvd., Stratford, CT 06615, USA.
URL: http://www.oriel.com

Parker Hannifin Corp., General Valve Division, 19 Gloria Lane, Fairfield, NJ 07007, USA.
URL: http://www.parker.com

Perkin Elmer Ltd., Post Office Lane, Beaconsfield, Buckinghamshire HP9 1QA, UK.
Tel: 01494 676161
URL: http://www.perkin-elmer.com

Pharmacia Biotech (Biochrom) Ltd., Unit 22, Cambridge Science Park, Milton Road, Cambridge CB4 0FJ, UK.
Tel: 01223 423723
Fax: 01223 420164
URL: http://www.biochrom.co.uk
Pharmacia and Upjohn Ltd., Davy Avenue, Knowlhill, Milton Keynes, Buckinghamshire MK5 8PH, UK.
Tel: 01908 661101
Fax: 01908 690091
URL: http://www.eu.pnu.com

Promega UK Ltd., Delta House, Chilworth Research Centre, Southampton SO16 7NS, UK.
Tel: 0800 378994
Fax: 0800 181037
URL: http://www.promega.com
Promega Corp., 2800 Woods Hollow Road, Madison, WI 53711-5399, USA.
Tel: 001 608 274 4330
Fax: 001 608 277 2516
URL: http://www.promega.com

Qiagen UK Ltd., Boundary Court, Gatwick Road, Crawley, West Sussex RH10 2AX, UK.
Tel: 01293 422911 Fax: 01293 422922
URL: http://www.qiagen.com
Qiagen Inc., 28159 Avenue Stanford, Valencia, CA 91355, USA.
Tel: 001 800 426 8157
Fax: 001 800 718 2056
URL: http://www.qiagen.com

Roche Diagnostics Ltd., Bell Lane, Lewes, East Sussex BN7 1LG, UK.
Tel: 01273 484644 Fax: 01273 480266
URL: http://www.roche.com
Roche Diagnostics Corp., 9115 Hague Road, PO Box 50457, Indianapolis, IN 46256, USA.
Tel: 001 317 845 2358
Fax: 001 317 576 2126
URL: http://www.roche.com
Roche Diagnostics GmbH, Sandhoferstrasse 116, 68305 Mannheim, Germany.
Tel: 0049 621 759 4747
Fax: 0049 621 759 4002
URL: http://www.roche.com

Schleicher and Schuell Inc., Keene, NH 03431A, USA.
Tel: 001 603 357 2398
URL: http://www.s-und-s.de

Shandon Scientific Ltd., 93-96 Chadwick Road, Astmoor, Runcorn, Cheshire WA7 1PR, UK. Tel: 01928 566611
URL: http://www.shandon.com

Sigma-Aldrich Co. Ltd., The Old Brickyard, New Road, Gillingham, Dorset XP8 4XT, UK.
Tel: 01747 822211 Fax: 01747 823779
URL: http://www.sigma-aldrich.com

Sigma-Aldrich Co. Ltd., Fancy Road, Poole, Dorset BH12 4QH, UK.
Tel: 01202 722114
Fax: 01202 715460
URL: http://www.sigma-aldrich.com
Sigma Chemical Co., PO Box 14508, St Louis, MO 63178, USA.
Tel: 001 314 771 5765
Fax: 001 314 771 5757
URL: http://www.sigma-aldrich.com

Stratagene Inc., 11011 North Torrey Pines Road, La Jolla, CA 92037, USA.
Tel: 001 858 535 5400
URL: http://www.stratagene.com
Stratagene Europe, Gebouw California, Hogehilweg 15, 1101 CB Amsterdam Zuidoost, The Netherlands.
Tel: 00800 9100 9100
URL: http://www.stratagene.com

Supelco, Supelco Park, Bellenfonte, PA 16823-0048, USA.
URL: http://www.sigma.sial.com

Sutter Instruments, 51 Digital Drive, Novato, CA 94949, USA.
URL: http://www.sutter.com

TefLabs, 9503 Capitol View Drive, Austin, TX 78747, USA.
URL: http://www.teflabs.com

United States Biochemical, PO Box 22400, Cleveland, OH 44122, USA.
Tel: 001 216 464 9277

VisiTech International Ltd., Sunderland Enterprise Park (East), Sunderland SR5 2TA, UK.
URL: http://www.visitech.co.uk

Index